D1378289

Large Ion Beams

Large Ion Beams

Fundamentals of Generation and Propagation

A. Theodore Forrester
Professor of Physics and Electrical Engineering
University of California at Los Angeles

WILEY

A Wiley-Interscience Publication

JOHN WILEY & SONS

New York • Chichester • Brisbane • Toronto • Singapore

Library of Congress Cataloging in Publication Data:

Forrester, A. Theodore.
   Large ion beams: fundamentals of generation and propagation/A.
Theodore Forrester.
        p.   cm.
   "A Wiley-Interscience publication."
   Bibliography: p.
   Includes index.
   ISBN 0-471-62557-4
   1. Ion bombardment.   I. Title.
QC702.7.B65F67   1987
530.4'1—dc19   87-24186
                                                      CIP

Printed in the United States of America
10 9 8 7 6 5 4 3 2 1

To Lloyd P. Smith. It was he who introduced me to the dynamics of charged particles, and I take this opportunity to thank him.

Preface

When students came to my laboratory in which ion beam research was conducted, they generally knew little about the type of physics that is essential to research on ion beam generation, extraction, acceleration, deceleration, and propagation. They might have studied electromagnetism but the chances are that they had never heard of the Child or Child–Langmuir equation. If they had studied plasma physics, they knew a great deal about instabilities in plasmas and wave propagation in plasmas but very little about the relationship between plasma parameters and available ion current density, and they rarely had the background for understanding the criteria for designing electrodes for extracting and accelerating an ion beam.

Many books have been written about ion sources and ion beams and some of them are good. However, I find it difficult to tell students who think they would like to do some work on ion sources or ion beams to start by reading any particular book. I tried to correct this by giving a series of lectures to my students, but the problem of instilling the fundamentals in this manner is that the lectures that are appropriate for the novice are too elementary for the seasoned members of the team.

I therefore put together a compilation of elementary discussions of phenomena that a student who wants to work on ion beams should be familiar with. As this material accumulated, it seemed worthwhile to organize and expand it into a book which would serve students of high perveance ion beams generally. As the book progressed, it became much more than a primer and I believe it will prove to be, as well, a valuable reference even to veterans in the field of ion source development.

The description of the various types of positive ion sources covered in Chapter 8 is, I believe, the most thorough categorization given anywhere, although radio frequency sources are slighted. I felt that the organization deserved the word *taxonomy*, although I refrained from classifying the sources into phyla, classes, genera, and species, as I might have. Other compilations have allowed function to guide the tabulation. For example, one reference classifies ion sources which are used in neutral beam lines as atom beam sources, in spite of the fact that the identical sources could be applied to propulsion or sputtering. In my book, the sources are categorized in terms of their configurations, and these are analyzed more thoroughly than elsewhere, so that, for the first time, as far as I am aware, the relationship between the various types of sources is developed. It is interesting to note the similarity between the periplasmatron and duoPIGatron sources, and the resemblance between a duoPIGatron and a hollow cathode Kaufman-type source is made clear for the first time, I believe.

The field of intense negative ion beam production is a rapidly developing field, which made the organization of the materials of Chapter 10 more difficult. Nevertheless, I believe the organization of that material will provide an important boost for anyone entering this unusually challenging area.

While the history of types of sources is close to complete, I am aware that I have not included all sources that have merit. I apologize to those whose sources have not been recognized here in spite of some advantageous features.

Finally, there are places in this book where I offer suggestions for interesting research directions. I hope that some of them prove to be meritorious. They are all directions I would pursue if I had the resources.

A. THEODORE FORRESTER
Los Angeles, California

Supplementary Preface:

The Menace of SDI

As the writing of this book was nearing completion, I attended the Negative Ion Symposium in Brookhaven* in order to bring up to date the chapter on the rapidly changing field of negative ion sources. I was chagrined to discover that the conference was sponsored by the Strategic Defense Initiative Office (SDIO) and that many of the participants were supported by the SDIO. I say chagrined because I had been devoting myself to speechmaking and writing condemning the Strategic Defense Initiative (SDI), or "Star Wars" program as a pursuit which increased the probability of a nuclear war between the superpowers. The thought that this book might be used as preparation for work that could produce such a holocaust was disturbing—and yet I could not abandon this project to which I had devoted a very large effort. I chose to resolve the moral dilemma by including this preface. It is hoped that it will be read, will be persuasive in exposing the menace of "Star Wars," and will cause some to eschew participation in this destabilizing and provocative development.

The menace arises from the fact that the Soviet Union views SDI, in fact must view SDI, as an attempt to develop the capability of coping with a response to a first strike by the United States, and not as a defense against a first strike by the Soviet Union. I want to develop that point and describe the responses which the Soviets are likely to make to our deployment of such a system.

Note first of all that a 99% effective SDI system would permit approximately 100 warheads to strike us even if the Soviet arsenal had no more strategic warheads than it now possesses. It is doubtful that we could recover from such devastation—

---

*Fourth International Symposium on the Production and Neutralization of Negative Ions and Beams, Brookhaven National Laboratory, October 27–31, 1986.

and defensive systems *never* approach that level of effectiveness. Consider, for example, antiaircraft weaponry. The Battle of Britain was won because the British were able to inflict losses of approximately 10% per raid on the Luftwaffe. Since then antiaircraft means have become much more effective and include missiles that actively seek their targets. However, evasive actions have also become much better and it is usually estimated that even today in a determined bomber strike no more than 10% of the attacking force could be brought down. That is effective against conventional bombing, but 10% is worthless when the nature of nuclear weapons is considered.

Can anyone seriously consider that for the first time in the history of warfare a perfect shield will prove feasible, especially when that shield will probably be the most complicated technological undertaking in the history of civilization? The proposals for the means to accomplish this task border on the ludicrous, and yet, since they do not violate any fundamental physical principles, I shall not claim that the task of making an effective shield against the present generation of ICBMs is impossible. What is not conceivable is that in the competition between defensive weapons and evasive actions available to the offense, including proliferation in numbers and resort to alternative delivery systems, the defensive system can always be close to 100% effective.

Another point is even more telling. There can be no such thing as an invulnerable satellite and, since the pursuit of SDI precludes an agreement to ban anti-satellite (ASAT) weapons, such weapons will be extremely efficient well before SDI can be deployed. Should the Soviets decide to strike they would simply take out the space units of the SDI system simultaneous with their launch. Thus SDI is simply not a defensive system at all.

There is, however, a context in which SDI is embraced by some sort of wild logic. There may be planners who feel that in the time required to develop and deploy our SDI system we could have every Soviet submarine followed by a killer sub, and by this and other means have the capability of launching a first strike so devastating that the Soviets would be able to retaliate with only a few hundred, rather than thousands, of warheads. With our SDI system on full alert, they may fantasize we could cope with these. It was made plain by Secretary of Defense Caspar Weinberger that we would never stand for the deployment of an SDI system by the Soviets, but he failed to make clear what we would do. Let us consider the steps the Soviet Union is likely to take should we attempt to deploy an SDI system:

1. They may multiply the number of their warheads to be able to overwhelm anything we put up. It has been estimated that they would multiply the number of strategic warheads by a factor of 5 for 10% of the cost of an SDI system. This would result in an unconstrained arms race.

2. They may shoot down the orbiting components of our SDI system. The minimum retaliation on our part would be destruction of Soviet satellites. The result would be a space war resulting in the elimination of all space systems such as surveillance, communications, and weather satellites. Would such a war lead to a nuclear exchange?

3. Fearing a U.S. first strike, they may go to a launch-on-warning response, which would allow no more than perhaps two minutes between the observation of a radar signal indicating an incoming missile and a launch. Accidental nuclear war will then become probable.

4. The fear of a first strike by the United States may lead them to launch a preemptive strike.

The horror of these probable responses is great. The chance that the Soviets will sit back and agree to limit the number of their missiles to whatever number we believe a proposed SDI system can cope with is negligible. Not only is any attempt to deploy an SDI system extremely destabilizing, but even the pursuit of SDI is damaging to our security. If we were not determined to develop SDI, we could have a comprehensive test ban treaty, a ban on ASAT weapons, a ban on all weapons in space, and we could take steps that would increase the stability of deterrence and eliminate the possibility of a nuclear winter.

My favorite scheme is the elimination of all MIRVed missiles, in fact all silos and submarine-launched missiles, in favor of small mobile missiles with single warheads of no more than 50 kilotons of TNT explosive yield. If each superpower were limited to, say, 1000 such missiles, the world would become a much safer place. Whether this, or one of many other directions, is the right one will require study by both sides but it seems overwhelmingly clear that our security depends on mutual carefully monitored agreements and that SDI, which precludes agreements, only makes the world a more dangerous place.

This essay is written to implore you not to enter into SDI-related research, to urge you not to take the attitude that you do the research and let the politicians make the decisions. Please consider the consequences of your own actions.

Contents

Large Ion Beams

Chapter 1

Introduction

Those who are interested in utilizing ion beams for many different applications—ion propulsion, neutral beam injection into magnetic fusion devices, sputtering, or ion implantation, for example—are faced with the task of understanding and dealing with the problems of generation, extraction, and propagation of the ion beam. In many cases, space charge and plasma phenomena dominate other considerations. Books on ion sources and ion beams have not given adequate emphasis to these areas or to ion source design fundamentals. Wilson and Brewer (1973), for example, emphasize specific application to ion implantation. Lawson's (1977) theoretical treatment touches some of the problems we cover, but it treats only beam propagation and does not deal with sources. Jahn's (1968) treatise on electric propulsion is more concerned with the collective phenomena involved in electrothermal and magnetohydrodynamic thrusters than with phenomena germane to large ion sources and ion beams. Stuhlinger's (1964) book on ion propulsion for space flight only begins to touch on aspects of ion beam generation that are essential to the innovative design and development of large ion sources. Brewer's (1970) volume on ion propulsion and Kaufman's (1974) survey of ion thruster technology do more in this direction, but neither touches, for example, on the material of Chapters 2 and 3 of this book, which I regard as essential for the novice and important for the pursuit of research or development of large ion sources. The book by Valyi (1977) is a valuable reference in which one can find an extraordinarily large collection of ion sources. However, none of these books appears to me appropriate for the novice, the young physicist or electrical engineer who has a background in electromagnetism, perhaps even plasma physics, and wants to work on ion beam generation.

For example, in Chapter 3 of this book relationships between ion source performance characteristics such as ion current density and neutral gas efflux and the plasma parameters such as electron density and temperature are developed. This material is lacking in the aforementioned books and, as far as I am aware, in sources other than periodicals or compilations of journal articles. In general, this book attempts to cover the generation, acceleration, and propagation of ion beams in such a way as to provide a solid basis for the student of the subject to understand ion sources and carry out constructive research on ion source improvement.

The problem areas to be surveyed are best illustrated by examining the generalized picture of an ion source seen in Fig. 1.1. A cathode held at a negative potential of the order of 100 volts relative to an anode is made to emit electrons into a region containing the gas or vapor to be ionized. Under the appropriate pressure conditions, a plasma fills the plasma generator region of the ion source and the primary electrons are accelerated to their full energy across a narrow sheath at the cathode. The electron flow is determined by the space charge limitations in this sheath or the emission limitations of the cathode. For a large source, the cathode may have to supply many hundreds, even thousands of amperes. Producing such currents for the desired pulse lengths, or for steady state operation, with the long lifetime always required is one of the most critical problems of ion sources and Chapter 7 is devoted to cathodes for this purpose.

The problem of efficiently using the primary electrons for ionization has been attacked mainly by the use of magnetic fields, in a variety of configurations (none of which is indicated in Fig. 1.1) to contain the electrons and increase the proba-

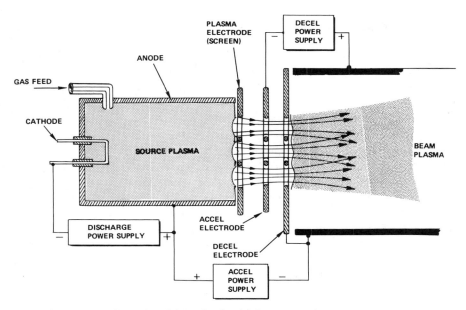

Figure 1.1. Schematic of an ion beam generating system.

bility of ionization taking place. The success in achieving the goal of high ion-iation efficiency by the plasma generator depends on the containment configuration utilized. Ions formed in the source plasma drift toward the various boundaries of the region containing the plasma. One of the subjects to be discussed is the design of the magnetic and electrostatic field configuration by which a large fraction of the generated ions is guided to the extraction area.

The ions that reach the plasma electrode are accelerated by an electrostatic grid structure. The energy of the extracted ions is determined by the voltage on the power supply labeled in Fig. 1.1 as the accel power supply. The actual supply voltage across the gap between the plasma and accel electrodes is the sum of the voltage on the accel and decel power supplies. After an acceleration to a higher energy than required, the ion beam must undergo a deceleration to its final energy. The existence of a potential minimum between the source plasma and the ion beam is required because the ion beam region, for sufficiently high currents, is occupied by a plasma, and a barrier to inhibit electron flow back to the ion source is nec-essary. Ion flow in the electrode region is determined by a solution of Poisson's equation with appropriate boundary conditions. We shall examine the space charge considerations which are essential for the design of electrodes for ion beam ex-traction and deceleration.

An inefficient use of the gas not only leads to excessive pumping requirements but can provide serious problems with the extraction electrode system. Charge exchange between the neutral gas and ions in the accelerating region causes wide angle parasitic components of the beam, even for electrodes designed to produce a highly directed beam. It also causes ion bombardment of the accel electrode. The resulting sputtering limits electrode lifetime, and secondary electron emission can cause excessive heating of the plasma electrode and ion source walls. Consid-erations related to increasing the gas utilization efficiency are therefore covered in some detail.

The electrode system shown in Fig. 1.1 often lacks a material decel electrode. For reasons already given a deceleration of the ions is essential, but the deceler-ation may take place between the accel electrode and a virtual electrode at the boundary of the beam plasma region. In fact, we shall see in Chapter 5 that it is possible to make an accel–decel electrode system with only one electrode.

The problem of the beam propagation after it leaves the ion source and extrac-tion electrodes has many interesting aspects. The ion beam generally spreads more than can be explained by the range of angles at which the beam is launched. The nature of this spreading and its causes, as discussed in Chapter 6, contain important clues as to how such beam spreading might be minimized.

The physics common to every part of the problem is the physics of space charge phenomena. Mathematically, this implies solutions of Poisson's equation, which is given, in SI or mks units, by

$$\nabla^2 V = -\rho/\varepsilon_0 \qquad (1.1)$$

where V is the electrostatic potential, ρ is the charge density, and ε_0 the permittivity of free space. The escape of electrons from an emitter, the drift of ions toward a plasma boundary, the acceleration of an ion beam across a gap, and the spreading of an intense ion beam are all space charge or plasma problems. The next two chapters are therefore devoted to a study of these phenomena in idealized formulations of real problems. Most of the ideas presented in these chapters were developed long ago. They are given here because they are needed in a useful form and are no longer a part of the studies of the physicist or engineer, even those whose specialty is plasma physics. Some of the material is presented here for the first time.

Finally, I want to remark on the illustrative problems presented at the end of some chapters. I strongly urge the reader to work these problems. Not only do they illustrate the material and let students test their level of understanding, but in many cases they make important points which extend the text coverage. For this reason I have chosen to include not merely answers but a complete set of solutions to the problems as an appendix.

Chapter 2

Collisionless Space Charge Phenomena

2.1 Scope

Space charge phenomena are phenomena associated with the flow of charged particles in sufficient numbers that their density substantially changes the potential in the region through which they flow. In this chapter we consider ion and electron flows in various configurations, but in each case the motion of a particle is determined solely by the electrostatic field produced by the electrodes and the collective action of the other charged particles. The discussion of the influence of binary collisions or other equilibration processes is deferred to subsequent sections of the book.

A typical collisionless space charge problem of interest for ion sources is that of determining electrode geometry and the potentials that must be applied to extract ions from a plasma, accelerate the ions to a desired energy with minimum electrode interception, and launch them as a beam with small angular spread. To make it possible to face a challenge of that magnitude, it is necessary to proceed through a sequence of steps of increasing complexity. It is helpful to start with the analysis of charged particle flow between infinite parallel planes.

2.2 Plane Parallel Electrodes (The Child Equation)

Consider the flow of charged particles between two parallel planes, such as the flow of electrons from a planar cathode to a planar anode. Let us idealize by supposing the planes to be infinite in extent and by assuming that the charged particles are emitted with zero velocity. For explicitness we assume the charged particles

5

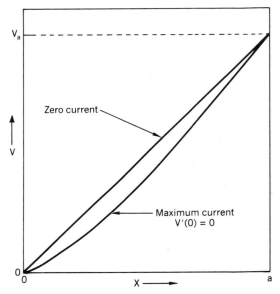

Figure 2.1. Potential distributions between parallel plane electrodes.

are electrons* for which $q = -e$. The results will be applicable to ions by reversing the sign of the potential. This problem was first solved by Child (1911) and later reworked by Langmuir (1913).

The nature of potential distributions between a zero potential emitter at $x = 0$ and a collector at $x = a$ held at a potential V_a is seen in Fig. 2.1. As the current flow is increased, the space charge causes a lowering of the curve until the electric field goes to zero at $x = 0$. For particles emitted with zero velocity, this defines the maximum current, and the solution corresponding to this case is of special interest.

For this one-dimensional situation, Poisson's equation (1.1) becomes

$$V''(x) = -\rho/\varepsilon_0, \qquad (2.1)$$

where $V''(x)$ is used for d^2V/dx^2. In Eq. (2.1) we may let

$$\rho = -J/v, \qquad (2.2)$$

where J is the magnitude of the current density and v is the particle velocity. Since

---

*Electrons, rather than particles with charge q (+ or −), are used to avoid having a velocity $\sqrt{-qV/M}$. The elimination of the minus sign under the radical is hardly a compelling reason, but then there is no reason not to. The application of the results to positively charged particles presents not the slightest difficulty.

the flow is unidirectional and steady, J is independent of x, and the velocity v is given by

$$v = (2eV/m)^{1/2}. \qquad (2.3)$$

When Eqs. (2.2) and (2.3) are inserted into Eq. (2.1), we obtain

$$V''(x) = (J/\varepsilon_0)(m/2eV)^{1/2}, \qquad (2.4)$$

to be solved with initial conditions

$$V' = 0 \quad \text{and} \quad V = 0 \quad \text{at } x = 0. \qquad (2.5)$$

The first integration of Eq. (2.4), simplified by multiplying the left side by $2V'dx$ and the right side by the equivalent $2\,dV$, leads to

$$(V')^2 = 4(J/\varepsilon_0)(mV/2e)^{1/2}. \qquad (2.6)$$

The integration constant has been set equal to zero to satisfy the conditions of Eq. (2.5). The variables V and x in Eq. (2.6) easily separate and the resulting equation can be integrated to yield

$$V = (m/2e)^{1/3}\,(J/\varepsilon_0)^{2/3}\,(3x/2)^{4/3}. \qquad (2.7)$$

Again, the conditions of Eq. (2.5) require that the integration constant be zero. Since $V = V_a$ at $x = a$, we obtain

$$J = (4\varepsilon_0/9)\,\sqrt{2e/m}\;V_a^{3/2}/a^2, \qquad (2.8)$$

the well-known Child equation,* in mks units. It is convenient to write this equation using a constant χ (chi for Child), such that

$$J = \chi V_a^{3/2}/a^2, \qquad (2.9a)$$

where

$$\chi = (4\varepsilon_0/9)(2e/m)^{1/2}. \qquad (2.9b)$$

For electrons, this constant is $2.334 \times 10^{-6}\ \text{A}/\text{V}^{3/2}$, and the value for singly charged ions is shown as a function of the atomic mass value in Fig. 2.2. It is helpful to note that the use of a in centimeters rather than meters in the Child equation is convenient and leads to a value of J in amperes per square centimeter.

---

*This equation is often called the Child–Langmuir equation, but Child's publication so clearly ante-dates Langmuir's that I have chosen to call it the Child equation.

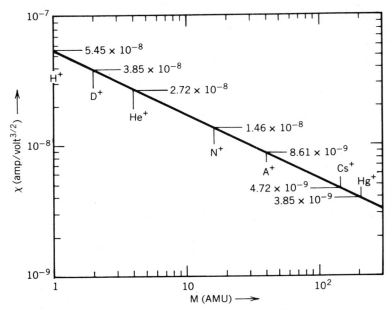

Figure 2.2. Child equation constant $\chi = (4\varepsilon_0/9)(2e/M)^{1/2}$ for singly charged ions.

It might be noted that ρ, given by Eq. (2.2), and therefore $V''(x)$, is infinite at the emitter surface. This has posed no difficulty since the infinity has proven to be integrable, but it may be conceptually bothersome. In a real situation, no such infinity exists because of a particle velocity spread. The degree to which the solution obtained here approximates the more realistic solution will be seen in Sec. 2.7.

2.3   Coaxial Cylinders

Before deriving the equation which is equivalent to Eq. (2.8) for coaxial cylinders and, in Sec. 2.4, concentric spheres, it is worth noting that the space charge limited electron current between an emitter at zero potential and an anode at potential V_a is always given by

$$I \propto V_a^{3/2} \tag{2.10}$$

as shown by Langmuir (1913). This can easily be seen as follows. The potential V at each point must be proportional to V_a on the collector if the normal component of the potential gradient at the emitter is maintained at zero. Poisson's equation (1.1) would then require that at each point

$$\rho \propto V_a. \tag{2.11}$$

Since the particle velocity given by Eq. (2.3) yields a velocity proportional to $\sqrt{V}$, we have, at each point,

$$v \propto V_a^{1/2}. \tag{2.12}$$

Then the current density at each point is

$$J = \rho v \propto V_a^{3/2}. \tag{2.13}$$

It follows that the total current satisfies Eq. (2.10) under space charge limited conditions, for any two-electrode geometry.

For coaxial cylinders V is a function only of the distance r from the axis and Poisson's equation can be written as

$$\frac{1}{r}\frac{d}{dr}\left(r\frac{dV}{dr}\right) = -\frac{\rho}{\varepsilon_0}. \tag{2.14}$$

In this case ρ is given by

$$\rho = -i/2\pi r v \tag{2.15}$$

where i is an electron current per unit length and v is the velocity satisfying Eq. (2.3). Combining Eqs. (2.14), (2.15), and (2.3) yields

$$rd^2V/dr^2 + dV/dr = (i/2\pi\varepsilon_0)\sqrt{m/2eV}. \tag{2.16}$$

A closed form solution to this equation appears not to have been found but Langmuir (1913) expressed the solution as

$$i = (8\pi\varepsilon_0/9)\sqrt{2e/m}\,V^{3/2}/r\beta^2, \tag{2.17}$$

where β^2 is a function of the ratio of the radius r to the emitter radius r_0. In terms of the constant χ defined by Eq. (2.9b),

$$i = 2\pi\chi V^{3/2}/r\beta^2. \tag{2.18}$$

Langmuir and Blodgett (1923) substituted Eq. (2.17) into Eq. (2.16) and developed a series solution for β. Some of their tabulated values are given in Table 2.1. Note that the values of β^2 for converging cases ($r < r_0$) are tabulated as $(-\beta)^2$.

Equation (2.17) is readily solved for V and for any given i and the emitter radius r_0, $V(r)$ is readily found, as illustrated in the solution to Problem 2.5.

Table 2.1 Geometric Parameters for Space Charge Limited Flow between Cylinders and Spheres

$\dfrac{r}{r_0}$ or $\dfrac{r_0}{r}$	Cylindrical Parameter		Spherical Parameter	
	β^2	$(-\beta)^2$	α^2	$(-\alpha)^2$
1.0	0	0	0	0
1.1	0.00842	0.00980	0.0086	0.0096
1.2	0.02875	0.03849	0.0299	0.0372
1.3	0.05589	0.08504	0.0591	0.0809
1.4	0.08672	0.14856	0.0931	0.1396
1.5	0.11934	0.2282	0.1302	0.2118
1.6	0.1525	0.3233	0.1688	0.2968
1.8	0.2177	0.5572	0.248	0.502
2.0	0.2793	0.8454	0.326	0.750
2.5	0.4121	1.7792	0.509	1.531
3.0	0.5170	2.9814	0.669	2.512
4.0	0.6671	6.0601	0.934	4.968
5.0	0.7666	9.8887	1.141	7.976
6.0	0.8362	14.343	1.311	11.46
8.0	0.9253	24.805	1.575	19.62
10	0.9782	36.976	1.777	29.19
14	1.0352	65.352	2.073	51.86
20	1.0715	115.64	2.378	93.24
30	1.0908	214.42	2.713	178.2
50	1.0936	450.23	3.120	395.3
70	1.0878	721.43	3.380	663.3
100	1.0782	1,174.9	3.652	1,144
200	1.0562	2,946.1	4.166	3,270
500	1.0307	9,502.2	4.829	13,015

Key: r_0 = radius of emitter; r = radius of collector; β^2 and α^2 are for diverging cases, i.e., $r > r_0$; $(-\beta)^2$ and $(-\alpha)^2$ are for converging cases, i.e., $r < r_0$.

2.4   Concentric Spheres

It is easy to construct physical situations in which we have a space charge limited flow between parallel plane electrodes whose separation is so small compared to the lateral dimensions that the infinite parallel plane approximation is valid. The case of a cylindrical emitter and collector is very common, being represented, for example, by an emitting wire in the center of a cylindrical collector. It is not evident that the case of space charge limited flow between concentric spheres is a case of experimental interest because of the difficulty of making contact with the inner sphere. However, we shall see in Sec. 5.3 that we can work with a conical sector of a spherical geometry which retains the property of concentric spheres, that is, that the potential be a function of radius only. In addition, the field in the sheath around a negative spherical probe in a plasma will be governed by the equations developed here in the limit of initial ion velocities $\ll$ sheath potential.

What parallel planes, coaxial cylinders, and concentric spheres have in common is that the electric field lines and charged particle paths are rectilinear and coincident. For the potential between concentric spheres Poisson's equation can be written as

$$\frac{1}{r^2}\frac{d}{dr}\left(r^2\frac{dV}{dr}\right) = -\frac{\rho}{\varepsilon_0}, \tag{2.19}$$

with ρ given by

$$\rho = -I/4\pi r^2 v, \tag{2.20}$$

and v, as before, satisfying Eq. (2.3). Combining these equations yields

$$\frac{d}{dr}\left(r^2\frac{dV}{dr}\right) = \frac{I}{4\pi\varepsilon_0}\left(\frac{m}{2eV}\right)^{1/2}. \tag{2.21}$$

As in the case of coaxial cylinders, a closed form solution appears not to have been found, but Langmuir and Blodgett (1924) expressed their solution as

$$I = (16\pi\varepsilon_0/9)\sqrt{2e/m}\ V^{3/2}/\alpha^2, \tag{2.22}$$

where α^2 was found either in series form or as an integral, by substituting Eq. (2.12) into Eq. (2.20). A few values taken from their tabulations are shown in Table 2.1. Equation (2.22) can also be written as

$$I = 4\pi\chi V^{3/2}/\alpha^2. \tag{2.23}$$

For a given value of r_0, α^2 is a function of r, so that when I is specified, Eq. (2.23) gives the potential V as a function of r.

It is interesting to compare the emitter current densities in the three geometries we have covered. For the planar case we had

$$J = \chi V^{3/2} / x^2. \tag{2.24}$$

For the cylindrical case $J = i/2\pi r_0$ leads to

$$J = \chi V^{3/2} / r_0 r \beta^2, \tag{2.25}$$

and for the spherical case $J = I/4\pi r_0^2$ yields

$$J = \chi V^{3/2} / r_0^2 \alpha^2. \tag{2.26}$$

2.5   Momentum Balance Method

It is interesting to note that the Child equation can be derived by another route described by Forrester (1981) that does not use Poisson's equation and is simpler in that it avoids one integration. It is not important to rederive the Child equation but I proceed to do so as a means of introducing the method, which provides a tool that will be used for other problems as well, including plasma problems.

The potential $V(x)$ which satisfies the condition $V'(0) = 0$ is represented by the lower curve in Fig. 2.1. The replacement of the collector electrode at $x = a$ by one at a smaller value of x whose potential is $V(x)$ would leave the potential distribution between 0 and x unchanged. For any position of the collector electrode the net force on the emitter and collector together must be zero, simply by conservation of momentum. In this particular problem, with electrons emitted at zero velocity from an electrode at which $V' = 0$, there is no force on the emitter. We therefore require that the net force on the collector be zero. We equate the rate at which charged particles deliver momentum per unit area to the electric stress $\varepsilon_0 V'^2/2$ to obtain

$$(J/e)\, m(2eV/m)^{1/2} = \varepsilon_0 V'^2/2. \tag{2.27}$$

This can be recognized as equivalent to Eq. (2.6) and therefore leads to the Child equation (2.8).

2.6   Relativisitic Child Equation

Space charge is not usually important in ion beams at relativistic energies, because of the enormous power which the beam would have to carry, but relativistic corrections can be required for an analysis of electron flow. At any rate, the correction is so simply made with the technique of Sec. 2.5 that we do not need any compelling justification for its inclusion. Instead of mv, the relativistic case requires the use of a momentum

$$p = mv \left[1 - (v/c)^2\right]^{-1/2}, \tag{2.28}$$

where v at the collector plane can be obtained by setting the kinetic energy equal to the decrease in potential energy:

$$mc^2 \left(\left[1 - (v/c)^2\right]^{-1/2} - 1\right) = eV. \tag{2.29}$$

This leads to

$$\left[1 - (v/c)^2\right]^{-1/2} = 1 + (eV/mc^2), \tag{2.30}$$

from which we obtain

$$v = (2eV/m)^{1/2} \left[1 + (eV/2mc^2)\right]^{1/2} \left[1 + (eV/mc^2)\right]^{-1}. \tag{2.31}$$

Multiply the momentum per particle, as given by Eqs. (2.28), (2.30), and (2.31), by the particle flow rate per unit area J/e, and equate this to the electrostatic stress on the collector to satisfy the conservation of momentum criterion discussed in Sec. 2.5. We obtain

$$(Jm/e)(2eV/m)^{1/2} \left[1 + (eV/2mc^2)\right]^{1/2} = \varepsilon_0 V'^2/2. \tag{2.32}$$

If we solve for the potential gradient V' and separate variables, we obtain

$$V^{-1/4} \left[1 + (eV/2mc^2)\right]^{-1/4} dV = 2(J/\varepsilon_0)^{1/2} (m/2e)^{1/4} dx. \tag{2.33}$$

Let $eV/2mc^2 = s$ to obtain

$$\int_0^s \left[t(1 + t)\right]^{-1/4} dt = 2(J/\varepsilon_0)^{1/2} (m/2e)^{1/4} (e/2mc^2)^{3/4} x. \tag{2.34}$$

Although we have not been able to solve this integral generally, the assumption of small s enables us to convert the integrand to $t^{-1/4} [1 - (t/4)]$ and obtain

$$\int_0^s \left[t(1 + t)\right]^{-1/4} dt = \frac{4}{3} s^{3/4} \left(1 - \frac{3}{28} s + \cdots\right). \tag{2.35}$$

If we ignore higher order terms in s, Eqs. (2.34) and (2.23) yield

$$J = \left[(4\varepsilon_0/9) \sqrt{2e/m} \, V^{3/2}/x^2\right] \left[1 - \frac{3}{28} (eV/mc^2)\right], \tag{2.36}$$

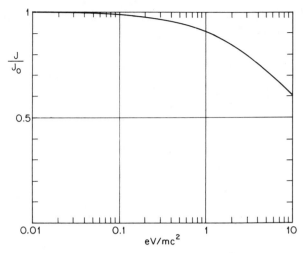

Figure 2.3. Ratio of the space charge limited current to the value obtained from the nonrelativistic Child equation.

which reduces to Eq. (2.8) when $(eV/mc^2) \ll 1$. The exact solution of Eq. (2.34) can be written as

$$J/J_0 = (9/16)\, s^{-3/2} \left\{ \int_0^s [t(1 + t)]^{-1/4}\, dt \right\}^2 \qquad (2.37)$$

where J_0 is the Child current given by Eq. (2.8). The results of a numerical integration are shown in Fig. 2.3 out to values of $2s = eV/mc^2 = 10$. These results are equivalent to the equations derived by Jory and Trivelpiece (1969), starting with Poisson's equation.

2.7   Sheath with Nonzero Initial Velocities

Consider now the problem of space charge flow between parallel planes when the condition $dV/dx = 0$ at the emitter is retained, but in which the emitted particles are all given an initial velocity v_0 perpendicular to the emitter surface. This is a special case of the space charge problems to be considered in Sec. 2.8, but it is one of special interest and more easily handled separately. It might at first seem to be an artificial problem since this is not a limiting flow situation as in the Child case. That is, particles with an initial energy could move to the collector in spite of a retarding field at the emitter. The condition $V' = 0$ at a boundary where injected particles have a directed velocity arises when an electrode is at a negative potential relative to a plasma. The plasma becomes the emitter of positive ions with a directed energy and the positively charged layer between the plasma, where $V' \approx 0$, and the electrode is commonly called a sheath. Real sheaths are more complicated than the idealized problem to be solved in this section, in that the ions have a spread in velocities, and electrons penetrate the sheath out to potentials of the order of kT_e/e. However, the term *sheath* is defined here as the layer between two equipotential surfaces when the requirement of a field intensity very close to zero is placed on one of them. This definition includes the region between two planes when the Child boundary conditions given in Eq. (2.5) are satisfied.

The differential equation to be solved is Eq. (2.4), but with initial conditions

$$V = V_0 \quad \text{and} \quad V' = 0 \quad \text{at } x = 0, \tag{2.38}$$

where $V_0 = mv_0^2/2e$.

There are convenient units for this problem. Let us use a dimensionless potential

$$\eta = V/V_0 \tag{2.39}$$

and a dimensionless distance

$$\xi = x/x_0, \tag{2.40}$$

where

$$x_0 = (2/3)\,\sqrt{\varepsilon_0/J}\,(2e/m)^{1/4}\,V_0^{3/4} \tag{2.41a}$$

or

$$x_0 = \sqrt{\chi/J}\,V_0^{3/4} \tag{2.41b}$$

is the Child acceleration distance corresponding to J and V_0. The constant χ is defined by Eq. (2.9b). In terms of the new variables η and ξ, Eq. (2.4) becomes

$$\eta'' = (4/9)\,\eta^{-1/2}, \tag{2.42}$$

with initial conditions

$$\eta' = 0 \quad \text{and} \quad \eta = 1 \quad \text{at } \xi = 0, \tag{2.43}$$

where the primes on η indicate differentiation with respect to ξ. The first integration yields

$$(\eta')^2 = (16/9)\,(\sqrt{\eta} - 1) \tag{2.44}$$

for the given initial conditions. This leads to

$$\int_1^\eta \left(\sqrt{t} - 1\right)^{-1/2} dt = 4\xi/3. \tag{2.45}$$

The substitution of a new integration variable for $\left(\sqrt{t} - 1\right)^{1/2}$ leads to an easily integrable form and we obtain

$$\left(\sqrt{\eta} - 1\right)^{1/2} \left(\sqrt{\eta} + 2\right) = \xi. \tag{2.46}$$

Substitution from Eqs. (2.39), (2.40), and (2.41) yields

$$\left(\sqrt{V/V_0} - 1\right)^{1/2} \left(\sqrt{V/V_0} + 2\right) V_0^{3/4} = (3x/2)(m/2e)^{1/4} (J/\varepsilon_0)^{1/2}. \tag{2.47}$$

For $\sqrt{V/V_0} \gg 1$, it can be seen that this equation becomes identical to Eq. (2.7) and therefore to the Child equation (2.9). Equation (2.47) cannot be solved explicitly for V, but we can plot $V(x)$ for any given value of J, V_0, and e/m, or make a universal plot of $\eta(\xi)$ from Eq. (2.46). This plot is seen in Fig. 2.4.

It is useful to solve Eq. (2.47) for the sheath thickness a, the distance in which the potential goes to a specific value of V, and compare with the sheath thickness obtained by solving the Child equation for x, obtaining

$$a/a_1 = \left(\sqrt{\eta} - 1\right)^{1/2} \left(\sqrt{\eta} + 2\right) \eta^{-3/4}, \tag{2.48}$$

where a_1 is the Child sheath thickness calculated for the same collector voltage V. To compute the comparative sheath thickness for the same drop across the sheath,

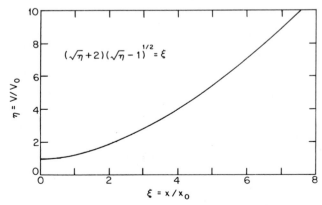

Figure 2.4. Potential variation through a sheath formed by charged particles of initial energy eV_0. The distance $x_0 = \sqrt{\chi/J}\, V_0^{3/4}$ is the Child equation distance corresponding to J and V_0.

it is necessary to compute the Child sheath thickness a_2 for a voltage $V - V_0$. We obtain

$$a/a_2 = (\sqrt{\eta} - 1)^{1/2} (\sqrt{\eta} + 2)(\eta - 1)^{-3/4}. \tag{2.49}$$

Equations (2.48) and (2.49) are plotted as curves (a) and (b), respectively, in Fig. 2.5. The thickness is seen to be significantly increased by initial ion energy even out to fairly high values of V/V_0. It is worth noting that

$$\frac{a}{a_2} \approx 1 + \frac{3}{2}\frac{1}{\sqrt{\eta}} \tag{2.50}$$

is a good approximation for values $\eta > 2$ and can also be used to represent a/a_1 for values of $\eta > 8$ with accuracy adequate for most applications.

In this analysis, we have considered the electrons or ions to be traveling from left to right in Fig. 2.1, or generally away from the zero potential gradient surface. The situation is exactly the same for ions traveling the other way, that is, decelerating toward a zero potential gradient boundary. One of the two sheaths of Problem 2.8 is a decelerating sheath satisfying this condition.

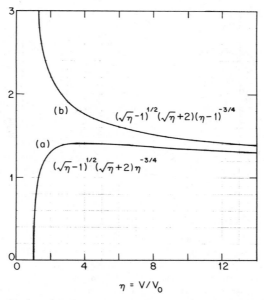

Figure 2.5. Ratio of the sheath thickness for an initial ion energy eV_0 to the Child equation sheath thickness for (a) the same energy at the collector and (b) the same voltage drop across the sheath.

2.8   Generalized Planar Space Charge Problem

The problem of charged particles injected into a region between two plane electrodes with a fixed energy eV_1, such as covered in Sec. 2.5, but without the requirement of zero field at one boundary has been covered by Salzberg and Haeff (1938) and Fay, Samuel, and Shockley (1938). In general there are four possible types of solutions, as illustrated in Fig. 2.6. As in Sec. 2.7, potentials are measured with respect to a point at which the electrons are at rest. Consider first the case shown in curve (a). In that case, the voltage on the right-hand electrode is negative, so that no electrons can reach it. All electrons are reflected at the point where the potential goes to zero and that point can be considered as a virtual emission-limited cathode emitting, as far as space charge is concerned, a current $2J$ where J is the current injected at $x = 0$. The solution in the region between the virtual cathode and $x = 0$ is a solution of Eq. (2.4) but with $V' \neq 0$ at the virtual emitter. In the region between the virtual emitter and the plane at $x = a$, V' is constant and the solution is that for which V' is continuous at the virtual emitter.

In curve (b) of Fig. 2.6, the potential goes to zero at some point with $V \geqslant 0$ at both boundaries. In general, this case would correspond to a reflection of a

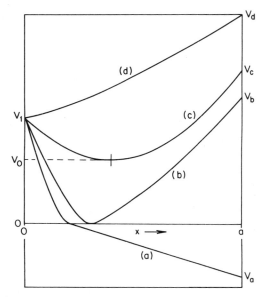

Figure 2.6. Types of potential distribution between parallel planes with unidirectional electrons of energy eV_1 injected on the left. Curve (a) is for the collector potential <0 so that all electrons are reflected. Curve (b) is for a collector voltage such that a fraction α ($0 < \alpha < 1$) of the injected electrons reaches the plane at $x = a$. In curve (c) a minimum potential $V_0 > 0$ exists between the two planes. Curve (d) represents the case of a monotonic acceleration between the two planes. A curve of type (d) can also have $V_d < V_1$, in which case the curve represents monotonic deceleration where all electrons reach the collector.

portion of the current, although in limiting cases none or all may be reflected. The case in which all would be reflected would be that for which $V_b = 0$ and in which $a \geq x_0$ where x_0 is the Child equation distance corresponding to the curent $2J$ and the voltage V_1 as illustrated in Problem 2.11. The turnaround point would occur at the value of x which satisfies the Child equation for a current $2J$ and a voltage V_1. A curve of type (b) could also correspond to no reflection if two Child solutions corresponding to the same current J exactly fit. In general, the type (b) curve would correspond to the transmission of a current αJ, $0 < \alpha < 1$, and the total solution consists of fitting into the available space and voltages a Child solution on the left corresponding to a current $(2 - \alpha)J$ and a Child solution on the right corresponding to a current αJ. It is the picture depicted in Problem 2.3, for example.

The case shown in curve (c) of Fig. 2.6 is that for which a minimum occurs between the two planes but with $V = V_0 > 0$ at the minimum. No charged particle reflection occurs and the solution corresponds to two solutions of the type given by Eq. (2.48) corresponding to the same current J and the same injection energy eV_0. In the limit of the minimum occurring at one of the two electrodes, this is precisely the case considered in Sec. 2.7. Finally, curve (d) illustrates the situations in which electrons are accelerated all the way and corresponds to a solution of Eq. (2.4) with V and dV/dx both positive at $x = 0$. Although the collector voltage is shown as larger than the injection voltage V_1, curves of the types (b), (c), and (d) can also have the collector voltage V_b, V_c, or V_d less than V_1.

The analyses of all cases are given in Salzberg and Haeff (1938) and Fay, Samuel, and Shockley (1938), but they are too extensive to warrant reproduction here. Curves of types (a) and (d) in Fig. 2.6 are not usually of interest since they represent a situation in which all or none of the current flows across the space. The case shown in curve (b) is easily solved using matched Child equations. The case illustrated by curve (c) has an interesting property that warrants further discussion because it leads to a bistable current flow in certain situations.

From the equations developed in Sec. 2.5, we can find the value of J corresponding to a value V_0 of the minimum. Let V_1 and V_2 be the insertion grid and collector potentials with $\eta_1 = V_1/V_0$ and $\eta_2 = V_2/V_0$. Let a be the grid-to-collector spacing with a_1 representing the distance between the injection grid and potential minimum. From Eq. (2.47), we obtain

$$\frac{a_1}{x_0} = (\sqrt{\eta_1} + 2)(\sqrt{\eta_1} - 1)^{1/2} \tag{2.51}$$

and

$$\frac{a - a_1}{x_0} = (\sqrt{\lambda\eta_1} + 2)(\sqrt{\lambda\eta} - 1)^{1/2}, \tag{2.52}$$

where $\lambda = V_2/V_1 = \eta_2/\eta_1$. Add Eqs. (2.51) and (2.52) to obtain

$$a = \left[(\sqrt{\eta_1} + 2)(\sqrt{\eta_1} - 1)^{1/2} + (\sqrt{\lambda\eta_1} + 2)(\sqrt{\lambda\eta_1} - 1)^{1/2} \right] x_0. \tag{2.53}$$

Insert the expression for x_0 given by Eq. (2.41b) with $V_0 = V_1/\eta_1$, and solve for the current density J, to obtain

$$J = \left[(\sqrt{\eta_1} + 2)(\sqrt{\eta_1} - 1)^{1/2} + (\sqrt{\lambda\eta_1} + 2)(\sqrt{\lambda\eta_1} - 1)^{1/2}\right] \eta_1^{-3/2} J_0, \quad (2.54)$$

where $J_0 = \chi V_1^{3/2}/a^2$ is the C-L current corresponding to the voltage V_1 and spacing a.

For simplicity, consider the case $V_2 = V_1$ for which $\lambda = 1$. Equation (2.54) becomes

$$J = 4(\sqrt{\eta_1} + 2)^2 (\sqrt{\eta_1} - 1) \eta_1^{-3/2} J_0. \quad (2.55)$$

In Fig. 2.7, J/J_0 is plotted as a function of $1/\eta_1 = V_0/V_1$ and it is to be noted that the current goes through a maximum at a value of V_0 called V_{0m}. For this particular case, $V_{0m} = V_1/4$, but in general V_{0m} will depend on $\lambda = V_2/V_1$, as seen in Fig. 2.8. Potential distributions corresponding to $0 < V_0 < V_{0m}$ can be seen to be inaccessible solutions. Imagine, for example, holding the voltages V_1 and V_2 fixed while the current density increases. As it does so, V_0 will decrease until $V_0 = V_{0m}$. Any further increase in the injected current must be accompanied by a reflection, implying a curve of the type (b) in Fig. 2.6. If the injected current is then decreased, the transmitted current will increase until the minimum reaches its furthest distance from the injection plane and more of the injected current is transmitted. For $V_2 = V_1$, the minimum would be at the midplane with $J = 4J_0$. Any futher decrease in injected current would cause a discontinuous change in the potential to a curve of type (c) in Fig. 2.6. For the $V_2 = V_1$ case, we can see from Fig. 2.7 that this would correspond to $V_0 = 0.75 V_1$.

Neither can the potential distributions corresponding to $V_0 < V_{0m}$ be reached

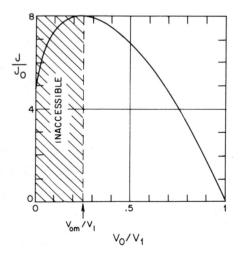

Figure 2.7. Current density as a function of the minimum potential V_0 for a common voltage V_1 on the insertion grid and collector. J_0 is the Child equation current corresponding to V_1 and the grid to collector spacing.

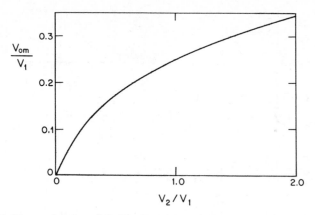

Figure 2.8. V_{0m}/V_1 as a function of V_2/V_1. V_{0m} is the minimum value of V_0 corresponding to the maximum current flow between planes at potentials V_1 and V_2.

by voltage variations. As an illustration, let us work out a particular numerical example. Suppose a current $J_1 = 0.1$ A/cm^2 of electrons is injected through a fine grid at a potential $V_1 = 100$ volts with respect to the starting point of the electrons. A plane collector is parallel to the grid and separated from it by a distance $a = 0.4$ cm. Find the J–V curve to the collector as it is raised from zero volts until saturation, and then as the voltage is lowered back to zero.

As long as current is reflected, we have two Child equation curves, one for current αJ_1 and the other for $J_1(2 - \alpha)$. The current $J_1(2 - \alpha)$ corresponds to a voltage $V_1 = 100$ volts, and the current αJ_1 to the applied voltage V_2. We have, using Eq. (2.9a),

$$J_1(2 - \alpha) = \chi_e V_1^{3/2}/a_1^2 \tag{2.56}$$

and

$$\alpha J_1 = \chi_e V_2^{3/2}/(a - a_1)^2, \tag{2.57}$$

where χ_e is 2.334×10^{-6} A/V$^{3/2}$, and a_1 is the distance from the grid to the virtual cathode ($V = 0$ plane). These equations readily lead to $J = (\alpha J_1)$ versus V_2 in the ascending portion of Fig. 2.9, as follows. First, determine a_1 corresponding to $\alpha = 0$ from Eq. (2.56). In this case we obtain $a = 0.108$ cm. Get the value of a_1 for which all current flows to the collector plane by setting $\alpha = 1$. In this case we obtain $a_1 = x_0 = 0.1528$. The voltage V_2 corresponding to the minimum value of a_1 is zero since no current flows past the virtual cathode. The voltage V_2 corresponding to $\alpha = 1$ is easily found from Eq. (2.57) to be 190 volts by inserting $a_1 = 0.1528$. Points in between are found by varying the parameter a_1 in the range $0.108 < a_1 < 0.1528$ to find $J = \alpha J_1$ and V_2, and are displayed on the ascending portion of Fig. 2.9. When the voltage is lowered the current will stay at the max-

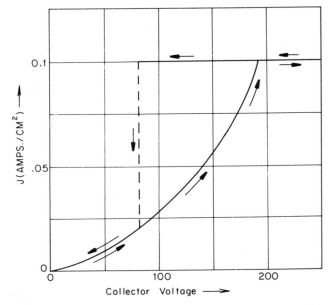

Figure 2.9. Calculated collector current as a function of collector voltage, for an injected current of 0.1 A/cm² at 100 volts and a spacing of 4 mm.

imum value until the value of V_0 drops to a point on the curve of Fig. 2.8. At this point the current which reaches the collector must drop discontinuously to the value corresponding to the rising portion of the curve, a shown in Fig. 2.9.

An adequate way of finding the voltage at which the discontinuity in current occurs is by choosing various values of V_2 which bracket the correct value. From V_2/V_1 find V_{0m} from Fig. 2.8. Using Fig. 2.4 the values of V_1/V_{0m} and V_2/V_{0m} lead to values of ξ_1 and ξ_2. For the injected current density, a value of x_0 and the collector to grid spacing $a = x_0(\xi_1 + \xi_2)$ is obtained. In this case, $V_2 = 85$ volts leads to $a = 4.07$ mm. A value $V_2 = 80$ volts yields $a = 3.98$ mm. The interpolated potential $V_2 = 81$ volts can be verified as giving the required type of potential distribution for a current of 0.1 A/cm² and a spacing of 4.00 mm. Between 81 and 190 volts, for the calculated case, the I–V curve shows hysteresis, that is, the collected current at a given voltage depends on the excursion of the voltage as it approaches the point.

Double valued solutions of the type seen in Fig. 2.9 were observed by Salzberg and Haeff (1938) and Fig. 2.10 presents examples taken from their paper. Situations in which ion beams actually encounter a similar bistable solution are not easily found. For one thing, an ion beam tends to eliminate potential maxima because of the trapping of electrons in such maxima. However, the possibility of such a bistable situation is a consideration in certain circumstances, especially in the deceleration of an ion beam for recovery of the ion beam power.

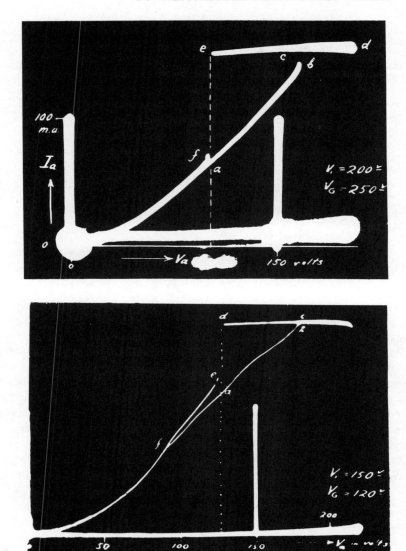

Figure 2.10. Collector characteristics observed by Salzberg and Haeff (1938) displaying the predicted hysteresis phenomenon.

2.9 Maxwellian Distribution of Emitted Velocities

Consider now the more realistic situation in which charged particles are emitted with a Maxwellian distribution. As in Sec. 2.2, consideration will be restricted to a planar vacuum diode, as illustrated in Fig. 2.11, with lateral dimensions so large

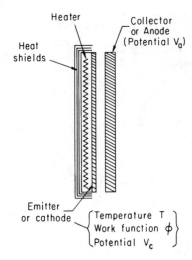

Heater

Collector
or Anode
(Potential V_a)

Heat
shields

Emitter
or cathode $\begin{cases} \text{Temperature T} \\ \text{Work function } \phi \\ \text{Potential } V_c \end{cases}$

Figure 2.11. Plane parallel vacuum diode.

compared to the interelectrode spacing that the current and voltage can be taken to be functions of a single spatial variable x. For electrons emitted with some initial energy it is clear that the maximum current for a given anode-to-cathode voltage is not the current corresponding to zero potential gradient at the emitter. Increases in emission beyond this value will create a retarding field at the emitter as shown in Fig. 2.12 but the type of potential distribution shown there corresponds to a greater current to the anode than one in which the minimum is at the cathode. Following Langmuir (1923), we consider this problem in terms of the flow of electrons of charge $-e$, but the results will apply equally well to ions emitted with a maxwellian distribution, as in the case of surface ionization.

The details of the analysis will be more easily followed if the nature of the solution is first described. A solution in closed form comparable to the Child equa-

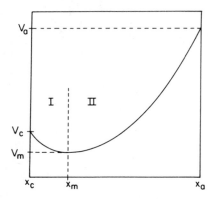

Figure 2.12. Nature of the potential distribution for space charge limited flow of electrons emitted with a spread in velocity. The emitter or cathode is at the left and the anode at the right.

tion appears to be unobtainable. However, it is feasible to obtain, numerically, a universal curve of potential as a function of distance in terms of dimensionless variables. Specifically, we define a dimensionless potential

$$\eta = \frac{V - V_m}{kT/e},$$ (2.58)

where V_m is the potential at the minimum and T is the emitter temperature, and a dimensionless length

$$\xi = 2\beta(x - x_m),$$ (2.59)

where

$$\beta^2 = (2\sqrt{\pi}/9)(J/\chi)(kT/e)^{-3/2},$$ (2.60)

with the Child equation parameter $\chi = (4e_0/9)\sqrt{2e/m}$ defined by Eq. (2.9b). The relationship between these variables will be shown to be given by*

$$\xi = \int_0^\eta (\epsilon^t - 1 \pm \epsilon^t \operatorname{erf} \sqrt{t} \mp 2\sqrt{t/\rho})^{-1/2} dt,$$ (2.61a,b)

where the function

$$\operatorname{erf} s = \frac{2}{\sqrt{\pi}} \int_0^s \exp(-t^2)\, dt$$ (2.62)

is the tabulated error function. The upper signs in Eqs. (2.61a and b) are valid in Region I of Fig. 2.12, that is, between the cathode and potential minimum, and the lower signs in Region II. These functions can be integrated numerically to yield a curve such as shown in Fig. 2.13, which can be used to obtain current voltage curves as follows. For a planar diode of given spacing a, choose a cathode temperature T and, for a given work function ϕ, use the Richardson equation

$$J_s = AT^2 \exp(-e\phi/kT),$$ (2.63)

discussed in detail, for example, by Jenkins and Trodden (1965), to obtain the emission current density J_s. Choose any value of current density J to the anode which is less than J_s. A maxwellian distribution is characterized by the fact that the current per unit area passing a retarding potential of magnitude V_r is given by

$$J = J_s \exp(-eV_r/kT).$$ (2.64)

*With ϵ representing permittivity and e the electronic charge, I use the symbol ϵ for the napierian base, that is, $\epsilon^t = \exp(t)$.

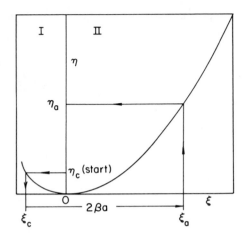

Figure 2.13. Illustration of the method of using $\eta(\xi)$ for obtaining $V(x)$ or $I(V)$ curves. (Curve is schematic only, not quantitative.)

For emitted electrons the potential barrier is $V_c - V_m$ so that

$$J/J_s = \exp(-\eta_c), \tag{2.65}$$

where $\eta_c = e(V_c - V_m)/kT$. Thus our starting conditions on J_s and J give

$$\eta_c = \ln(J_s/J). \tag{2.66}$$

From η_c and the $\eta(\xi)$ curve in Region I, ξ_c is easily obtained. Equation (2.60) gives β and the dimensionless cathode–anode distance is determined from Eq. (2.59) to be

$$\xi_a - \xi_c = 2\beta a, \tag{2.67}$$

where a is the cathode-to-anode spacing. When this distance is added to ξ_c, as seen in Fig. 2.13, the dimensionless anode position coordinate ξ_a is obtained. The corresponding value η_a of the normalized anode voltage is then obtained. The required anode–cathode voltage is then

$$V_a - V_c = (\eta_a - \eta_c)\,kT/e. \tag{2.68}$$

This procedure could be repeated for various choices of $J < J_s$ leading to $J(V)$ curves.

It is useful to note that $kT/e = 1$ volt when $T = 11,600$ K. Thus, if potentials are in volts, kT/e can be replaced by $(T/11,600)$.

To carry out the analysis leading to Eqs. (2.61a) and (2.61b), it is necessary to split the electrons into increments of equal x-directed emitted velocity v_0 and then integrate over v_0. In the planar diode we are not concerned about transverse velocities and when we use the term kinetic energy we mean the kinetic energy associated with x velocities.

The current due to particles with initial kinetic energies between eV_0 and $e(V_0 + dV_0)$ is the same as the decrease in the current produced by increasing a retarding potential from V_0 to $V_0 + dV_0$. From Eq. (2.64) we obtain

$$dJ = (J_s e/kT) \exp(-eV_0/kT) \, dV_0. \qquad (2.69)$$

But

$$eV_0 = mv_0^2/2 \qquad (2.70)$$

and

$$edV_0 = mv_0 dv_0, \qquad (2.71)$$

so that Eq. (2.69) becomes

$$dJ = (J_s mv_0/kT) \exp(-mv_0^2/2kT) \, dv_0. \qquad (2.72)$$

The region between the cathode and potential minimum, Region I of Fig. 2.12, contains electrons flowing in both the $+x$ and $-x$ directions. It is simpler to analyze Region II where all electrons have positive x velocities, and we proceed to do that first.

Each current increment contributes to the space charge at each point an amount

$$d\rho = -dJ/v, \qquad (2.73)$$

where v is the x component of velocity at an arbitrary value of x. The total space charge is obtained by integrating over all electrons which surmount the potential barrier at x_m. This gives

$$\rho = -(mJ_s/kT) \int_{v_m}^{\infty} (v_0/v) \exp(-mv_0^2/2kT) \, dv_0, \qquad (2.74)$$

where

$$v_m = \sqrt{2(V_c - V_m)e/m}. \qquad (2.75)$$

The velocity v is related to v_0 and potential V through

$$\tfrac{1}{2}mv^2 = \tfrac{1}{2}mv_0^2 + (V - V_c)e. \qquad (2.76)$$

At a given position, V is fixed and

$$mv \, dv = mv_0 dv_0. \tag{2.77}$$

Use Eqs. (2.76) and (2.77) to change the integration variable from v_0 to v obtaining

$$\rho = -(mJ_s/kT) \exp \left[e(V - V_c)/kT \right] \int_{v_1}^{\infty} \exp \left(-mv^2/2kT \right) dv \tag{2.78}$$

where

$$v_1 = \sqrt{2(V - V_m) e/m}. \tag{2.79}$$

But the current to the anode is

$$J = J_s \exp \left\{ -\left[e(V_c - V_m)/kT \right] \right\} \tag{2.80}$$

so that Eq. (2.78) becomes

$$\rho = -\left(\frac{mJ}{kT} \right) \exp \left[\frac{e(V - V_m)}{kT} \right] \int_{v_1}^{\infty} \exp \left(\frac{-mv^2}{2kT} \right) dv. \tag{2.81}$$

Let $mv^2/2kT = s^2$ obtaining

$$\rho = -J \sqrt{\frac{2m}{kT}} \, \epsilon^{\eta} \int_{\eta}^{\infty} \exp \left(-s^2 \right) ds, \tag{2.82}$$

where η is the dimensionless potential defined by Eq. (2.58). Since $d^2\eta/dx^2 = e/kT \, d^2V/dx^2$, Poisson's equation $d^2/V/dx^2 = -\rho/\varepsilon_0$ becomes

$$\frac{d^2\eta}{dx^2} = \left(\frac{eJ}{\varepsilon_0 kT} \right) \sqrt{\frac{2m}{kT}} \, \epsilon^{\eta} \int_{\eta}^{\infty} \exp \left(-s^2 \right) ds. \tag{2.83}$$

But the integral from $\sqrt{\eta}$ to infinity is the difference between the integral from zero to infinity and the integral from 0 to $\sqrt{\eta}$. Since the integral of $\exp \left(-s^2 \right)$ from 0 to ∞ is $\sqrt{\pi}/2$ we obtain

$$\int_{\sqrt{\eta}}^{\infty} \exp \left(-s^2 \right) ds = \frac{\sqrt{\pi}}{2} (1 - \text{erf} \sqrt{\eta}), \tag{2.84}$$

where erf $\sqrt{\eta}$ is defined by Eq. (2.62). The substitution of ξ for x, using Eqs. (2.59) and (2.60), simplifies Eq. (2.83) to

$$\eta'' = \tfrac{1}{2} \epsilon^{\eta} (1 - \text{erf } \sqrt{\eta}). \qquad (2.85)$$

This equation needs to be solved subject to $\eta = 0$ and $\eta' = 0$ at $\xi = 0$. Equation (2.85) can be integrated by the same technique used to integrate Eq. (2.41), obtaining

$$\eta'^2 = \int_0^{\eta} \epsilon^t (1 - \text{erf } \sqrt{t}) \, dt, \qquad (2.86)$$

which becomes

$$\eta'^2 = \epsilon^{\eta} - 1 - \int_0^{\eta} \epsilon^t \text{ erf } \sqrt{t} \, dt. \qquad (2.87)$$

An integration by parts gives

$$\int_0^{\eta} \epsilon^t \text{ erf } \sqrt{t} \, dt = \epsilon^{\eta} \text{ erf } \sqrt{\eta} - 2\sqrt{\frac{\eta}{\pi}}. \qquad (2.88)$$

Put this into Eq. (2.87), take the square root of both sides, and integrate, obtaining Eq. (2.61b), the desired Region II solution.

In Region I of Fig. 2.12 it is helpful to split ρ into two parts, part (a) due to electrons moving toward the anode and part (b) due to electrons moving toward the cathode. In (a), there are electrons for which

$$e(V_c - V) < \tfrac{1}{2} m v_0^2 < \infty, \qquad (2.89)$$

and in (b),

$$e(V_c - V) < \tfrac{1}{2} m v_0^2 < e(V_c - V_m). \qquad (2.90)$$

It will be left as an exercise for the reader to follow the procedure carried out for Region II [or consult Langmuir (1923)] and show that in Region I

$$\eta'' = \tfrac{1}{2} \epsilon^{\eta} (1 + \text{erf } \sqrt{\eta}) \qquad (2.91)$$

replaces Eq. (2.85), and that this equation solves to yield Eq. (2.61a).

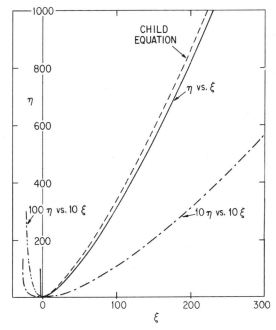

Figure 2.14. Potential as a function of distance in terms of dimensionless variables, as given by Eqs. (2.61a) and (2.61b).

The required plot of $\eta(\xi)$ is shown in Fig. 2.14, which can be used as shown in Fig. 2.13.

For large η and $\xi > 0$, Langmuir found a series expansion for $\xi(\eta)$. The use of the first two terms leads to

$$J = (4\varepsilon_0/9)\sqrt{2e/m}\,(V - V_m)^{3/2}(x - x_m)^{-2}(1 + 2.66\eta^{-1/2}). \quad (2.92)$$

The potential dip $V_c - V_m$ cannot be very large compared to kT/e so that η large and $V \gg V_m$ are nearly equivalent statements. Further, $V \gg V_m$ implies an $x \gg x_m$ so that for large η we can write

$$J = J_0(1 + 2.66/\sqrt{\eta}), \quad (2.93)$$

where J_0 is the Child equation current. It is the fact that $J \to J_0$ as $\eta \to \infty$ that really justifies the use of the Child equation but it is worth noting that η must be extremely large if the Child equation is to be regarded as accurate. For example, if $\eta = 500$, the current predicted by Eq. (2.93) is 12% higher than J_0. For a cathode at 2320 K, $kT/e = 0.2$ volt and $\eta = 500$ corresponds to a voltage of 100 volts.

Another indication of the level of error made in using the Child equation is shown in Fig. 2.14. In terms of the variables η and ξ used in this section the Child equation becomes

$$\eta = (\tfrac{9}{8}\sqrt{\pi})^{2/3}\xi^{4/3}, \tag{2.94}$$

which is plotted as the broken line.

2.10   Double Sheath with Zero Emitted Velocities

All of the space charge problems considered to this point have been unipolar problems, that is, have involved only a single species of charged particle. In the situation in which a space charge limited electron emitter is held negative relative to a plasma, a sheath is formed in which ions from the plasma are accelerated toward the electron emitter and electrons from the emitter are accelerated toward the plasma. The potential variation as shown in Fig. 2.15 must have zero slope on the left for a space charge limited flow of electrons emitted with zero velocity, and zero slope on the right to match the field in the plasma. It is called a double sheath because electron space charge dominates in one half and ion space charge in the other. Let us first solve the problem idealized to the case of electrons and ions both emitted with zero velocity at two fixed plane boundaries with a potential difference v_a between them.

Langmuir (1929) solves this problem as a solution of Poisson's equation. It is also possible to avoid one integration by using the momentum balance technique of Sec. 2.3 as done by Forrester (1981). Note first that we can regard any plane between $x = 0$ and $x = a$ as the source of the ions, but to duplicate the situation that exists between the plane and the electron emitter, we would have to think of

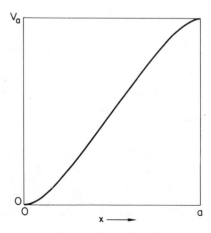

Figure 2.15. Potential distribution for space charge limited bipolar flow.

the plane as a source of ions with a velocity $\sqrt{2e(V_a - V)/M}$, where V is the potential at the plane and M is the mass of an ion. These ions would, of course, produce a recoil. The total momentum balance equation would be obtained by setting the force on the cathode equal to the force on this imaginary plane,

$$\frac{J_i}{e} M \sqrt{\frac{2eVa}{M}} = \frac{J_e}{e} m \sqrt{\frac{2eV}{m}} + \frac{J_i}{e} M \sqrt{\frac{2e(V_a - V)}{M}} - \frac{\varepsilon_0 V'^2}{2}, \qquad (2.95)$$

where the last term is the electrostatic stress on the ion emitting plate.

By setting $V = V_a$ where $V' = 0$, we readily obtain

$$J_i/J_e = \sqrt{m/M}. \qquad (2.96)$$

Eliminate J_i from Eq. (2.95) by substitution from Eq. (2.96), let $\eta = V/V_a$ and $\xi = x/a$ to obtain

$$\eta'^2 = (16/9)(J_e/J_0)(\sqrt{1 - \eta} - 1 + \sqrt{\eta}), \qquad (2.97)$$

where

$$J_0 = (4\varepsilon_0/9)(2e/m)^{1/2} V_a^{3/2}/a^2 \qquad (2.98)$$

is the Child current corresponding to the voltage V_a and spacing a. Since $\eta = 0$ at $\xi = 0$, Eq. (2.97) becomes

$$\int_0^\eta (\sqrt{1 - s} - 1 + \sqrt{s})^{-1/2} \, ds = \frac{4}{3} \sqrt{\frac{J_e}{J_0}} \xi. \qquad (2.99)$$

From $\eta = 1$ at $\xi = 1$, we obtain

$$\frac{J_e}{J_0} = \frac{9}{16} \left[\int_0^1 (\sqrt{1 - s} - 1 + \sqrt{s})^{-1/2} \, ds \right]^2. \qquad (2.100)$$

This integral can readily be evaluated numerically to yield

$$J_e/J_0 = 1.8651.^* \qquad (2.101)$$

It can be seen from the nature of the integral in Eq. (2.99) that the potential is symmetric about the midplane. The symmetry leads to the conclusion that the ion

---

*Langmuir (1929) obtained the value 1.8605. The value 1.8651 is the result of a better calculation than he could have done without modern calculators. Levine (1984) obtained this result analytically.

flow will be greater than the Child space charge limited ion flow by the same factor (1.8651). The square root of this number gives

$$a/a_0 = 1.3657 \qquad (2.102)$$

where a_0 is the C–L distance corresponding to the same unipolar flow of either J_e or J_i.

In the real case of a sheath between a cathode and a plasma, the plasma side is penetrated by plasma electrons and the ions reach the sheath boundary with a significant energy. We shall see in Sec. 3.7 the difference made by these departures from the idealization of this section.

Larger effects than given by Eq. (2.101) can be obtained by introducing the ions at a plane for which $0 < x < a$. Since this is difficult to effect experimentally, this general case is not analyzed here but is covered by Langmuir (1929).

2.11 Space Charge Effects in Beams—Perveance and Poissance

Consider now charged particles forming a beam, which we define as a flow of particles with lateral boundaries. In this section we consider the idealized situation of identical ions, all of the same injected energy eV, initially traveling parallel to each other and perpendicular to an equipotential injection plane. We shall consider long, narrow beams and beams of circular cross section in a region bounded by equipotential surfaces. We first examine beams of low current, by which we mean here beams whose space charge produces a potential variation through the beam much less than the potential, and then the case of beams whose space charge effects are capable of producing beam stalling. Ions traveling parallel to each other produce magnetic fields which yield attractive forces. In ion beams of high energy the magnetic forces approach the size of the electrostatic repulsive forces. This becomes an important consideration in the very high power, very short-pulse ion diodes being studied for light ion inertial confinement fusion [Miller (1982), and VanDevender (1986)]. In the long-pulse and continuous operation ion diodes we consider here, the conditions for magnetic pinch do not apply. Let us consider the various electrostatic cases separately.

A. Long Narrow Beams, Poissance < 1

We consider the beam to be injected into a region through a slit of width a and length b, with $b \gg a$. As a result of lateral fields due to space charge, the ions will not continue to travel parallel to each other but will diverge at increasing angles, relative to the plane of symmetry with increasing distance z. If the spreading angle is small, the lateral field E can be obtained from Gauss' law.

The charge density in the beam is given by

$$\rho = J(M/2eV)^{1/2}, \tag{2.103}$$

where V is the potential relative to a point at which an ion originates, J is the injected current density, and M is the ion mass. If we take J as uniform, we readily obtain

$$E_x = (Jx/\varepsilon_0)(M/2eV)^{1/2} \tag{2.104}$$

when x, the distance from the median plane, is less than $a/2$. We have implicitly assumed that the walls are at the potential V and are close to the edge of the beam. For this case, there is no large potential drop between the beam edge and the walls, and the potential variation ΔV across the beam is assumed to be small compared to V. Otherwise, the buildup of potential would retard the ions, increase charge density, and increase E_x. The wall position is subject to experimental control and the value of ΔV can be calculated to find the range of currents consistent with the assumption of small ΔV.

We integrate Eq. (2.75) from $x = 0$ to $x = a/2$ to obtain

$$\Delta V = (Ja^2/8\varepsilon_0)(M/2eV)^{1/2}. \tag{2.105}$$

Dividing by V yields

$$\Delta V/V = J/18J_0 \tag{2.106}$$

where

$$J_0 = \chi V^{3/2}/a^2 \tag{2.107}$$

with

$$\chi = (4\varepsilon_0/9)(2e/M)^{1/2} \tag{2.108}$$

is the Child equation current which can be drawn across a gap equal to the beam width, as given by Eqs. (2.9a) and (2.9b). We recognize that for values of J up to the order of J_0 the assumption of $\Delta V \ll V$ is quite good.

Since the value of E_x outside the beam is constant, it is easy to obtain the potential drop $\Delta V'$ between the beam and the bounding equipotential surface. This yields

$$\frac{\Delta V'}{V} = \frac{J}{9J_0}\left(\frac{a'}{a} - 1\right), \tag{2.109}$$

where a' is the width of the channel through which the long slit beam is to propagate. The previous conclusion that the slowing down of the beam could be neglected out to values of J of the order of J_0 would be good out to channel widths no more than twice the beam widths.

If we assume that the potential in the beam is the potential V at which it is injected, so that Eq. (2.104) is correct, we obtain the acceleration eE_x/M of an ion at the edge of the beam by setting $x = a/2$. This yields

$$d^2x/dt^2 = (Ja/2\varepsilon_0)(e/2M)^{1/2}V^{-1/2}. \qquad (2.110)$$

Because J may vary across the beam, and will certainly decrease as the beam expands, it is preferable to replace J with the current $I = Jab$, to obtain

$$d^2x/dt^2 = (I/2b\varepsilon_0)(e/2M)^{1/2}V^{-1/2}. \qquad (2.111)$$

The acceleration at the beam edge remains constant as the beam expands so that we can easily calculate $x(t)$ for an ion which starts at $x = a/2$ and $z = 0$. The result is

$$x = (a/2) + (I/4b\varepsilon_0)(e/2MV)^{1/2}t^2. \qquad (2.112)$$

The time t can be eliminated by using $z = vt$ with $v = \sqrt{2eV/M}$. We obtain

$$x/a = (1/2) + (I/18I_0)(z/a)^2, \qquad (2.113)$$

where $I_0 = J_0ab$ is given by

$$I_0 = \chi V^{3/2}(b/a). \qquad (2.114)$$

Because space charge is proportional to $I/V^{3/2}$, the effects are usually described by the quantity

$$P = I/V^{3/2}, \qquad (2.115)$$

called the perveance. However, space charge effects are related not only to perveance but to the ion mass as well. An ion beam perveance often is given by the equivalent electron beam perveance by multiplying P by $\sqrt{M/m}$. This still fails to include the geometric factor b/a and has the disadvantage of its dimensionality so that one must remember that 2.334 equivalent electron micropervs, where a microperv is 10^{-6} A/V$^{3/2}$, produces very large space charge effects in ion beams for which $b/a = 1$. Another terminology sometimes used is aspect ratio, meaning the square of the ratio of the width of the aperture from which the ions are extracted to the distance between the accel and plasma electrodes of Fig. 1.1. For electrode systems in which the accelerating distances are no more than three times the slit width this quantity is made somewhat meaningless by various effects, such as the bowing of the equipotential surfaces and the effect of the accel–decel voltage ratio,

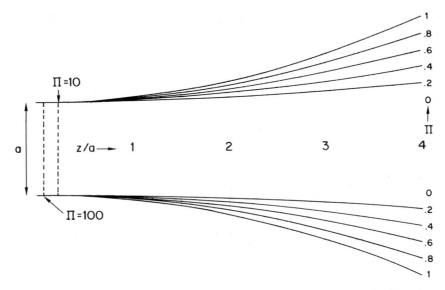

Figure 2.16. Space charge spreading of long slit ion beams for low poissance and stalling distance for high poissance. Poissance $\Pi = (I/\chi V^{3/2})(a/b)$, where b is length, a is width, and χ is the Child constant $(4\varepsilon_0/9)\sqrt{2e/M}$ plotted in Fig. 2.2.

so that the quantity I/I_0, seen in Eq. (2.113) to describe space charge spreading, is not uniquely determined by the aspect ratio. In fact the quantity I/I_0 itself, which might be called an equivalent aspect ratio, is the best parameter to use and I call this quantity the *poissance** Π. The definition $\Pi = I/I_0$, with I_0 given by Eq. (2.114) leads to

$$\Pi = (I/V^{3/2})/(\chi b/a), \qquad (2.116)$$

so that the poissance is seen to be a normalized perveance. In terms of this parameter Eq. (2.92) becomes

$$(x/a) = (1/2) + (1/18)\,\Pi(z/a)^2, \qquad (2.117)$$

and is plotted in Fig. 2.16 for various values of Π. We see that the poissance must be $\ll 1$ if the spreading of an unneutralized beam is to be small.

If the beam were initially uniform, the lateral force on an ion would be proportional to distance from the median plane, as seen in Eq. (2.104). This would lead to an expansion which would maintain beam uniformity. It is worthwhile noting again that the conclusions of this section, such as this one, are based on the assumption of an ion beam with no spread in velocity or direction and containing

---

*Pronounced pwahsance, never poysance.

no other charged particles such as electrons or slow ions. More realistic situations are considered later.

B. Axially Symmetric Beams, Poissance <1

Now consider a beam of circular cross section. As before, we consider the case in which all ions have identical velocity and initially are traveling parallel to the axis and for which space charge effects are small enough that the potential variation through the beam can be neglected. If we consider sufficiently small divergence angles, we can take the field as purely radial and use Gauss' law to obtain

$$E_r = (Jr/2\varepsilon_0)(2eV/M)^{-1/2}. \tag{2.118}$$

To determine how large the current can be, consistent with the low space charge assumption, we integrate E_r from 0 to the beam radius r_0, obtaining

$$\Delta V = (Jr_0^2/4\varepsilon_0)(2eV/M)^{-1/2}, \tag{2.119}$$

which leads to

$$\Delta V/V = (I/4\pi\varepsilon_0)(2e/M)^{-1/2}V^{-3/2}, \tag{2.120}$$

where we have set $J\pi r_0^2 = I$. We can write

$$\Delta V/V = \Pi/9\pi, \tag{2.121}$$

where Π is the poissance defined by Eq. (2.116) with $b = a$, that is,

$$\Pi = I/\chi V^{3/2}. \tag{2.122}$$

As in the case of long narrow beams, the treatment of the beam as having negligible potential variations is justified for poissance values up to the order of unity.

The distance to the surrounding, presumably cylindrical wall must also be considered. The field outside the beam falls off as $1/r$ and the integration readily yields

$$\frac{\Delta V'}{V} = (\Pi/4.5\pi)\ln(R/r_0), \tag{2.123}$$

where $\Delta V'$ is the potential drop between the beam and the bounding surface of radius R. This situation is much more favorable than for the long slit beams and until R/r_0 becomes quite large we can consider $\Delta V'/V \ll 1$ for values of $\Pi \leq 1$.

From Eq. (2.118) we can obtain the radial acceleration $(e/M) E_r$ of an ion at the beam edge. We obtain

$$d^2r/dt^2 = (I/2\pi\varepsilon_0 r)(e/2MV)^{1/2}. \tag{2.124}$$

Integrating once yields

$$(dr/dt)^2 = (I/\pi\varepsilon_0)(e/2MV)^{1/2} \ln(r/r_0), \tag{2.125}$$

where the integration constant has been chosen to make $dr/dt = 0$ at $r = r_0$. By replacing t with $z = t\sqrt{2eV/M}$, we obtain

$$dr/dz = (I/2\pi\varepsilon_0)^{1/2}(M/2e)^{1/4}V^{-3/4} \ln^{1/2}(r/r_0), \tag{2.126}$$

leading to

$$\int_1^{r/r_0} \ln^{-1/2} s\, ds = \left(\frac{2\Pi}{9\pi}\right)^{1/2}\frac{z}{r_0}. \tag{2.127}$$

If we let $\ln^{1/2} s = u$, we obtain

$$\int_0^{\ln^{1/2}(r/r_0)} \exp(u^2)\, du = \left(\frac{\Pi}{18\pi}\right)^{1/2}\frac{z}{r_0}. \tag{2.128}$$

Although the use of poissance has been introduced here, this equation is equivalent to one given by Fowler and Gibson (1934). The integral in Eq. (2.128) is easily related to Dawson's integral

$$F(x) = \exp(-x^2)\int_0^x \exp(t^2)\, dt, \tag{2.129}$$

tabulated, for example, in Abramowitz and Stegun (1964). In terms of this function the integral in Eq. (2.128) is $(r_0/r) F\sqrt{\ln(r/r_0)}$.

The calculated beam spreading is shown in Fig. 2.17 for two values of poissance Π and is easily obtained from the curve for $\Pi = 1$ for any value of Π by multiplying the z coordinate by a factor of $\Pi^{-1/2}$.

We shall note in the next section that Smith and Hartman (1940) solved this problem for the general case where the slowing of the ions due to potential variations are taken into account.

C. High Current Beams

Although the potential variation through an ion beam of unit poissance is small, as seen in Eqs. (2.106) and (2.121), it is not possible to transmit beams whose

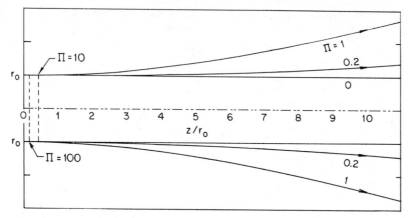

Figure 2.17. Space charge spreading of a circular ion beam for low poissance and stalling distance for high poissance. Poissance $\Pi = I/\chi V^{3/2}$.

poissance is much greater than unity without space charge neutralization of the beam. As the current is increased, the increased potential within the beam causes space charge to increase, causing a beam divergence greater than predicted by Eq. (2.128). However, the rapid increase in space charge density with current must lead to an actual stopping and reflection of the beam and this will happen discontinuously in a manner closely related to the phenomena described in Sec. 2.8. That is, the potential will fall continuously as the injected current is increased until a critical potential minimum greater than zero is reached. Any further increase in the injected current will lead to ion reflection and a decrease in the transmitted current.

The distance z_0 to the reflection plane will be the Child equation distance corresponding to a current density $2J$ and the voltage V:

$$z_0 = (2/3)(\varepsilon_0/2J)^{1/2}(2e/M)^{1/4}V^{3/4}. \tag{2.130}$$

For slit beams, this can be written as

$$z_0/a = (J_0/2J)^{1/2}, \tag{2.131}$$

where J_0 is defined by Eq. (2.98) and a is the beam width. In terms of the poissance

$$z_0/a = (2\Pi)^{-1/2}. \tag{2.132}$$

This stalling distance is shown for two values of Π in Fig. 2.16.

For circular beams, let $J = I/\pi r_0^2$ in Eq. (2.129). Then set $I/\chi V^{3/2} = \Pi$ to obtain

$$z_0/r_0 = (\pi/2\Pi)^{1/2} \qquad (2.133)$$

for the stalling distance displayed in Fig. 2.17 for $\Pi = 10$ and $\Pi = 100$.

The reflection surface is seen in Fig. 2.16 as if it were a plane. When the distance to this surface is small compared to the beam diameter, we can expect it to be planar over the central portion of the beam. Near the beam edge, however, we can expect the potential to be less positive than at the center, and the stalling surface together with ion trajectories can be expected to appear as shown in Fig. 2.18.

The transition cases between low poissance where the potential through the beam is essentially unchanged by space charge effects and that case in which reflection occurs was studied by Smith and Hartman (1940). Their work is not reproduced here, but a few facts would be interesting additions to the presentation given in this section. For a cylindrical beam of poissance 1 in a chamber whose diameter is 10 times the beam diameter, the beam would diverge to a given radius in 0.91 times the distance shown for $\Pi = 1$ in Fig. 2.16. For that case Eq. (2.102) indicates a potential drop between the beam and the wall which is 16.3% of the voltage V. A significant effect has to be expected. It is interesting that the correction is as small as it is justifying the use of Fig. 2.17 for values of Π as large as 1, as previously estimated.

It is feasible to overcome the beam spreading or stalling as described in this section by neutralization of the space charge by particles of the opposite sign. In fact, it is difficult to avoid trapping slow ions in an electron beam, or slow electrons

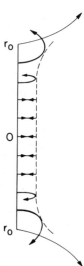

Figure 2.18. Expected ion trajectories for an unneutralized high poissance ion beam.

in an ion beam. These trapped particles reduce the well depth and therefore the lateral fields exerted on the beam particles. The type of beam spreading encountered in space charge neutralized ion beams is discussed in Chapter 6.

Problems

Section 2.2

2.1 A cesium diode is made up of two parallel plane electrodes separated by 0.05 mm. One of the electrodes is heated so that cesium comes off ionized. What is the space charge limited current density when an accelerating voltage of 10 volts exists between the two surfaces? Assume zero initial ion velocity.

2.2 A plane electron emitter is separated from a parallel plane anode by a distance of 1 mm. If the maximum power that can be dissipated at the anode is 100 W/cm^2, what is the maximum voltage that can be used under space charge limited flow conditions? What is the corresponding current density? Assume zero emission velocity.

2.3 A broad argon ion (A$^+$) beam with an energy of 100 eV per ion and a current density of 3 mA/cm^2 passes through a fine transparent grid. The other boundary of this region is a plane parallel electrode (the collector) at the same potential as the grid and 10 cm away. How much current can reach the collector if the region contains no space charge neutralization? Hint: Some of the current must turn back. This means that the potential goes to zero between the two 100 volt electrodes. The solution involves two back-to-back Child equations. Make approximations where valid and helpful.

2.4 Suppose a current density J to be made of a fraction α of ions of mass M_1 and a fraction $(1 - \alpha)$ of ions of mass M_2. What is the effective mass M_e that can be used in the Child equation?

Section 2.3

2.5 Plot $v(r)$ between a space charge limited cylindrical electron emitter of 2 cm radius and a coaxial collector 1 cm in radius held at 100 volts. Assume zero initial electron velocities. Compare this with the potential $V(x)$ that would exist between parallel plane electrodes separated by 1 cm.

Section 2.4

2.6 Suppose that a spherical probe of radius 1 mm immersed in a plasma draws a current of 1 mA of He$^+$ ions when held at a potential of 100 volts. Assume the ions start from rest at a spherical surface where $dv/dr = 0$, and that no electrons penetrate the region between this surface and the probe. Find the radius of the surface from which the ions get accelerated.

Section 2.5

2.7 In the derivation given in Sec. 2.5, the system is assumed to be isolated, since the net force is taken as zero. Does the fact that there are electric power leads attached to these plates and that electric power is being converted to thermal energy inside the diode invalidate this assumption? Explain.

Section 2.6

2.8 Find the space charge limited current density between an electron emitter and a parallel plane anode 50 cm away if the node is 5×10^6 volts relative to the cathode.

Section 2.7

2.9 In a collisionless plasma, it can be shown that most of the ions reach the plasma boundary with an energy close to $0.85kT_e$, where T_e is the electron temperature. Consider this number to be an energy of 4 eV for a particular plasma in which 50 mA/cm² of H^+ ions reach the boundary. Find the sheath thickness at a planar electrode near the boundary if the potential of the electrode is 50 volts negative relative to the center of the plasma where the ions have zero energy. What sheath thickness would be calculated if the ions were considered to reach the boundary with zero energy, and the potential drop across the sheath level to 50 volts, so that ions reach the collector with the same energy? What would you obtain if the initial energy was zero but the potential drop was the same as for the situation presented, that is, 46 volts?

2.10 The ion extraction system shown in Fig. 2.19 consists of a fine 75% transparent mesh facing an argon plasma. When the plasma is held at +1000 volts and the grid at −100 volts, the plasma boundary is 3 mm from the grid. The A^+ ion beam which goes through the grid is neutralized some distance downstream from the grid by a copious zero potential electron emitter which establishes the beam as a plasma at zero potential. Assume that the A^+ ion energy at the plasma boundary is so small compared to the 1000 volts across the acceleration gap that the Child–Langmuir equation is adequate for that region.
  (a)   What is the grid current per unit area?
  (b)   What is the ion beam current, downsteam of the grid, per unit area?
  (c)   Draw the $V(x)$ plot on graph paper, to scale, from the source plasma through the grid and into the ion beam plasma at $V = 0$. Treat the grid as an infinitely fine planar sheet.

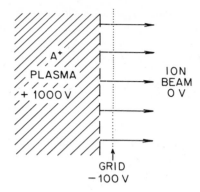

Figure 2.19. Ion extraction system for Problem 2.10.

Section 2.8

2.11 Suppose that a current of 9 mA/cm² of 100 volt electrons is injected through a fine mesh toward a plane collector 5 mm away held at a potential $V = 0^+$ (i.e., slightly greater than zero). Show that two solutions are possible and that one of them corresponds to the total reflection of the electrons and the other to zero reflection. (Voltages are given relative to a point where the electrons have zero kinetic energy.)

2.12 Suppose that a current of 30 mA/cm² of 100 volt electrons is injected into the diode of Problem 2.9. Plot $J(V)$, where V is the voltage on the collector, as V goes from 0 to 100 volts and back to zero again.

Section 2.9

2.13 A plane emitter at 2320 K produces a saturated current density $J_s = 0.5$ A/cm². If this emitter is the cathode of a planar diode with a spacing of 1 mm, plot $V(x)$ between the cathode and anode when a current of 0.05 A/cm² flows across the diode.

2.14 For the diode of Problem 2.13, plot $J(V)$ as the current density increases from zero to J_s. On the same curve plot $J(V)$ predicted by the Child equation.

2.15 If the emitter of Problem 2.1 emits 20 mA/cm² of Cs^+ ions at a temperature of 1000 K, find the space charge limited current corresponding to a 10 volt acceleration between emitter and collector. Compare with the result of Problem 2.1.

Section 2.10

2.16 The cathode of a mercury discharge runs 100 volts negative with respect to the plasma. The current of Hg^+ ions to the cathode is 2 mA/cm². Treat the ions as having zero velocity at the plasma boundary, and the electrons as emitted with zero velocity, and find:

(a)   The space charge limited emission from the cathode.

(b)   The thickness of the cathode sheath under space charge limited conditions.

Section 2.11

2.17   It is desired to send a beam of 10^4 eV Hg^+ ions 2 cm in diameter through a hole 0.5 cm in diameter 20 cm away. If there is no space charge neutralization, what is the maximum current in the beam such that all of the ions get through the hole? At what point on the axis would the ions have to be aimed?

Chapter 3

Collisionless Plasmas

3.1 Scope

In almost all intense ion sources the ions originate in a plasma.* It is important, therefore, to study the properties of this medium as a source of ions, including, for example, the relationship of ion current density to plasma parameters and the energy distribution to be expected in an ion beam originating in a plasma. Wave propagation, one of the primary subjects of most books on plasmas, is given only a brief summary and the subject of instabilities is not covered here at all.

3.2 Formation of a Plasma

Consider, following Langmuir (1929), the region between two parallel plane electrodes, one at $x = 0$ at a potential $V = 0$ and the other at $x = a$ and potential $V = -V_a$. Let us suppose that ions are generated at a uniform rate per unit volume g throughout the space between the electrodes. Whenever an ion is formed an electron is released but, for the moment, ignore the contribution of the electrons to the space charge. As long as the potential varies monotonically the electrons clear out so rapidly compared to the ions that this is a valid assumption. We shall also make the assumption that ions are formed at rest. The latter assumption is justified by the relatively low energy of neutral atoms and the very small amount of kinetic energy communicated to an ion in a collision with an electron.

As the ionization rate g is increased, the potential variation will change from the linear variation for $g = 0$ to a distribution in which $dV/dx = 0$ at $x = 0$, as

*The exception is the subject of Chapter 9.

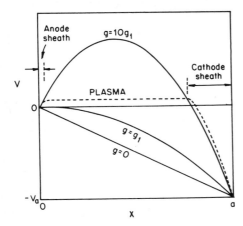

Figure 3.1. Potential distributions between parallel plane electrodes for uniform ion generation.

seen in Fig. 3.1. The value of g for the latter case is termed g_1. A further increase in g must produce a potential maximum as shown for the $g = 10g_1$ curve, but for this case the assumption that electrons will clear out so fast as to have a negligible effect on the potential becomes unjustified.

In fact electrons become trapped and a plasma must form. We shall return to this point later. Let us first find the critical value g_1 that marks the transition between a condition in which ions and electrons are swept out as they are formed, and the formation of a plasma.

It is convenient to treat this space charge problem by the momentum balance technique of Sec. 2.3. Since there is no current and no field at the positive electrode, there is no force on this electrode. We require, therefore, that an electrode placed at x with a potential $V(x)$ have momentum delivered at a rate exactly equal to the electrostatic force per unit area $\varepsilon_0 V'^2/2$ pulling the surface inward. The current which arrives at x coming from a range dx_1 at x_1,

$$dJ = g_1 e \, dx_1, \tag{3.1}$$

is made of particles with velocity

$$v = \sqrt{2e(V_1 - V)/M}, \tag{3.2}$$

where V_1 is the potential at x_1. Since dJ/e is the number of ions per second we obtain the momentum per second by multiplying this quantity by Mv and integrating over x_1:

$$\frac{\varepsilon_0 V'^2}{2} = g_1 \sqrt{2eM} \int_0^x \sqrt{V_1 - V} \, dx_1,$$

or

$$V'^2 = \frac{2g_1}{\varepsilon_0} \sqrt{2eM} \int_0^x \sqrt{V_1 - V} \, dx_1. \tag{3.3}$$

The simplest possible curve of the type shown for $g = g_1$ in Fig. 3.1 is given by

$$V = -V_a x^2/a^2 \tag{3.4}$$

and this is readily shown to be the solution by substituting in Eq. (3.3). We obtain

$$\frac{4V_a^2 x^2}{a^4} = \frac{2g_1}{\varepsilon_0 a} \sqrt{2eMV_a} \int_0^x \sqrt{x^2 - x_1^2} \, dx_1. \tag{3.5}$$

A change in variable $x_1 = x \sin \theta$ gives

$$\int_0^x \sqrt{x^2 - x_1^2} \, dx_1 = x^2 \int_0^{\pi/2} \cos^2 \theta \, d\theta = \frac{x^2 \pi}{4}.$$

When we put this in Eq. (3.5) we find that Eq. (3.4) is the solution providing that

$$g_1 = (4\varepsilon_0/\pi) \sqrt{2/eM} \, V_a^{3/2}/a^3. \tag{3.6}$$

Some features of this solution are interesting. The current which flows to the negative electrode,

$$J_1 = g_1 ea, \tag{3.7}$$

is then

$$J_1 = (4\varepsilon_0/\pi) \sqrt{2e/M} \, V_a^{3/2}/a^2, \tag{3.8}$$

which is seen to be

$$J_1 = (9/\pi) J_0 = 2.865 J_0, \tag{3.9}$$

where J_0 is the Child equation current corresponding to the voltage V_a and spacing a. The charge density, which, by Poisson's equation, is given by $-\varepsilon_0 V''$, is then

$$\rho = 2\varepsilon_0 V_a/a^2, \tag{3.10}$$

independent of x.

When $g > g_1$ the assumptions made lead to a solution of the type shown for $g = 10g_1$ in Fig. 3.1. The solution can be found by matching two solutions of the form given by Eq. (3.4) to obtain a parabolic solution with a maximum between $x = 0$ and $x = a$. The actual equation for the $g > g_1$ case is

$$V = -V_a(x/a)\left[1 - (g/g_1)^{2/3}(1 - x/a)\right],\tag{3.11}$$

but this is not a potential that would be observed. Whenever an ion is formed a slow electron is released. Such electrons would be trapped in a potential maximum, invalidating the assumption made concerning the neglect of electron space charge. In fact the filling in of the maximum by electrons will lead to a region in which the electron and ion space charges are approximately equal and the potential approximately constant, as shown by the dashed line in Fig. 3.1. The medium containing approximately equal densities of ion and electron space charge, called a plasma, will assume a potential positive enough with respect to the positive electrode so that electrons will leave at the same rate they are formed. If the potential drop across the region called the anode sheath were to go to zero, then electrons would clear out so fast compared to the ions a maximum would develop. By the formation of a nearly field-free region slightly positive relative to the anode, as seen in Fig. 3.1, the discharge contrives to find an appropriate equilibrium. The condition for the formation of a plasma is that g be greater than g_1. The phenomena governing the drift of ions to the plasma boundaries will be examined in some detail in Secs. 3.4 and 3.5.

3.3   Debye Shielding Distance

By measuring the $I(V)$ characteristics of small probes inserted into plasmas, Langmuir (1925) was able to determine that the distribution of energy among electrons in a plasma is maxwellian under a wide variety of conditions, even when the rapid thermalization cannot be understood on the basis of collisions among the particles.* Accordingly, the density n_e of electrons within a plasma can be related to potential V through the Boltzmann equation

$$n_e = n_0 \exp\left(eV/kT\right),\tag{3.12}$$

where n_0 is the density deep in the plasma where we take $V = 0$. The temperature T is in kelvins and k is the Boltzmann constant.

For simplicity consider a one dimensional potential variation such as might exist close to the surface of a planar electrode immersed in a plasma. We write Poisson's equation as

$$d^2V/dx^2 = -e(n_i - n_e)/\varepsilon_0,\tag{3.13}$$

assuming only singly charged ions. Take the potential V to be zero deep inside the

---

*The anomalously rapid maxwellianization of the electrons is known as the Langmuir paradox.

plasma where $n_i = n_e = n_0$. Suppose that n_i is insensitive to variations in potential and that n_e is given by Eq. (3.12). We obtain

$$d^2V/dx^2 = -en_0[1 - \exp(eV/kT)]/\varepsilon_0. \qquad (3.14)$$

For $eV/kT \ll 1$ we may let $\exp(eV/kT) = 1 + (eV/kT)$ to obtain the linearized equation

$$d^2V/dx^2 = (n_0e^2/\varepsilon_0kT)V, \qquad (3.15)$$

for which the solutions are

$$V = \exp\left[\pm(x/\lambda_D)\right], \qquad (3.16)$$

where

$$\lambda_D = (\varepsilon_0kT/n_0e^2)^{1/2} \qquad (3.17)$$

is called the Debye shielding length.

For the usual case of plasma dimensions $\gg \lambda_D$, the positive sign in Eq. (3.16) must be disregarded.

To find the potential around a point charge q in a plasma it would be necessary to write Eq. (3.14) in spherical coordinates and, subject to the same linearization with $eV/kT \ll 1$, obtain

$$V = (q/4\pi\varepsilon_0r) \exp(-r/\lambda_D) \qquad (3.18)$$

instead of the usual vacuum expression $(q/4\pi\varepsilon_0r)$.

We see that the distance λ_D defines the distance over which a steady state potential can have a large influence. In most laboratory plasmas it is a distance of 0.01–1 mm but can have any size. The most important criterion for a collection of charged particles to be termed a plasma is that the Debye length be small compared to the extent of the plasma. In Sec. 3.2 we implicitly assumed that electrons were cold enough that the anode sheath, which must be of the order of the Debye length, was small. If $\lambda_D > a$, then curves such as shown for $g = 10g_1$ in Fig. 3.1 could exist.

It is convenient to write Eq. (3.17) as

$$\lambda_D = \left[\frac{kT/e}{n_0(e/\varepsilon_0)}\right]^{1/2} \qquad (3.19)$$

since kT/e is the electron temperature in electron-volts (eV), the commonly used unit for plasmas, and $(e/\varepsilon_0) = 1.8 \times 10^{-8}$ volt-meter is a commonly occurring combination of the two constants. Note that in these units n_0 is in electrons per cubic meter and λ_D is in meters. In a set of units often used in the laboratory

$$\lambda_D(cm) = 745\left[T(eV)/n_0(cm^{-3})\right]^{1/2}. \tag{3.20}$$

For intense ion source plasmas T is approximately 5 eV and n_0 of the order of 10^{12} cm^{-3} yielding a λ_D of the order of 2×10^{-3} cm.

3.4 Bohm Criterion

The picture of a plasma developed in the last two sections is one in which we have an approximately equal density of ions and electrons. The electrons have a max-wellian distribution and can therefore unambiguously be assigned a temperature. We might imagine, at first, that the ions are also characterized by a nearly max-wellian distribution so that they too have a temperature and we would suppose that interactions between electrons and ions would cause these temperatures to be of the same order of magnitude. Even if ion temperatures were substantially larger, the electron flux through any area in the plasma would be so large compared to the ion flux that the condition that ions and electrons escape at approximately equal rates forces the plasma to be positive relative to the walls. In this case all ions that reach the plasma boundary are accelerated away from the plasma. The conse-quence is that the ions in the plasma, up to some collision distance inside the plasma, are all moving outward, that is, toward the nearest boundary. This is inconsistent with the idea of a maxwellian distribution.

Our picture then becomes one in which the ions, at least as the boundary is approached, are all drifting outward through a maxwellian cloud of electrons of density approximately equal to the ion density. We imagine the potential within the plasma to be substantially constant, as seen in Fig. 3.2. Let us suppose that the average outward velocity of the ions at the surface where the plasmas joins the sheath is v_0 and, for simplicity, suppose that all of the ions have this same directed velocity at this plane. It is easy to show that Poisson's equation (3.13) requires a minimum ion energy at this boundary, as follows.

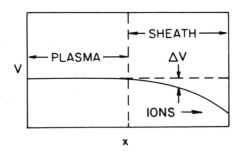

Figure 3.2. Potential variation at a plasma-sheath boundary.

The electrons penetrate the sheath according to the Boltzmann equation (3.12). The ion density in the sheath will vary according to

$$en_i = J_i/v \tag{3.21}$$

where J_i is the ion current density and v the normal component of the ion velocity in the sheath. Since $n_i = n_e$ in the plasma, call n_p the common density at the plasma to sheath transition. We can write

$$n_i = n_p v_0/v, \tag{3.22}$$

where v_0 is the value of v at the boundary. If $eV_0 = \frac{1}{2}Mv_0^2$ is the ion energy associated with the velocity v_0, then

$$n_i = n_p \sqrt{V_0/(V_0 + \Delta V)}, \tag{3.23}$$

where ΔV is the potential drop illustrated in Fig. 3.2. For very small ΔV, that is, for $\Delta V \ll V_0$ and $\Delta V \ll kT/e$, we have

$$n_i = n_p(1 - \Delta V/2V_0) \tag{3.24}$$

and

$$n_e = n_p(1 - e\Delta V/kT), \tag{3.25}$$

so that Poisson's equation becomes

$$d^2V/dx^2 = -(en_p/\varepsilon_0)[(e/kT) - (1/2V_0)]\Delta V \tag{3.26}$$

for very small ΔV.

If the term $(1/2V_0)$ is larger than (e/kT), then $d^2V/dx^2 > 0$, which means that the solution in the sheath must curve upward. It is clear from Fig. 3.2 that this is not a possible connection to the sheath consistent with our assumptions. We must require that $(1/2V_0)$ be less than (e/kT) or that

$$V_0 > kT/2e. \tag{3.27}$$

Although the requirement of a minimum ion energy at the plasma boundary was clearly pointed out by Tonks and Langmuir (1929), the simplified derivation given above is due to Bohm (1949, Chap. 3) and Eq. (3.27) has become known as the Bohm criterion. It is clear that the plasma potential cannot be perfectly flat. Equation (3.27) requires that within the plasma there be a sufficient field accelerating ions toward the boundary to provide the ions with a velocity greater than $\sqrt{kT/M}$ when it reaches the sheath boundary. The analysis of Sec. 3.5 shows how this presheath acceleration occurs and is consistent with the idea that the plasma contains nearly equal densities of electrons and ions.

If the directed ion energy at the plasma boundary is equal to the right hand side of the inequality (3.27), the current density would be

$$J_i = n_p e \sqrt{kT/M}, \tag{3.28}$$

where n_p is the ion density at the plasma–sheath transition. We shall see in Sec. 3.7 how this compares with the current calculated on the basis of the analyses of Secs. 3.5–3.7.

The analysis of this section leading to the condition (3.27) presupposed a maxwellian distribution of electrons. Plasmas which are maintained by the injection of electrons of energy much greater than kT have a small percentage, normally $< 10\%$, of primary electrons, that is, electrons with energies close to the injected energy. Masek (1969) showed that this increased the directed ion energy at the plasma boundary by a factor $n_i(0)/n_m(0)$, where $n_i(0)$ is the ion density, or total electron density, and n_m is the density of maxwellian electrons both taken at the plasma center. For the indicated small percentage of primaries, this increases the Bohm energy by no more than 10%. This slight a correction to its value can usually be ignored.

3.5 General Theory

Before we start the analysis a few general comments about the nature of the plasmas may be helpful. Since ions must reach the boundary of the plasma, that is, the edge of the volume in which $n_i \approx n_e$, with an average energy greater than $kT/2$, we must expect a variation in potential through the plasma of approximately kT/e, where T is the electron temperature. Since the plasma electron density varies as $\exp(eV/kT)$, we may expect a variation in electron density by a factor of 2 to 3. The existence of maxwellianized electrons right up to the boundary of the plasma requires that the plasma ride at a high enough potential relative to the boundary so that virtually all electrons are reflected; otherwise the maxwellian distribution would be truncated. This comes about, as demonstrated in Sec. 3.8, because the random electron flux in the plasma is so large compared to the ion current that unless the plasma rides at a potential several times kT/e relative to the wall, electrons will escape at a much greater rate than ions.

When fast primary electrons from the cathode enter the plasma they have an energy per electron an order of magnitude greater than kT. The density of these electrons in the plasma is insensitive to the relatively small potential variation through the plasma. If the ionization is produced by primary electrons, even though they constitute a very small fraction of the electron density, then the ionization rate will be uniform, if the density of neutral atoms is uniform. If, on the other hand, the main interaction of the primaries is with the maxwellianized electrons, with most of the ionization produced by the high energy tail of the latter group, then the rate of ionization per unit volume will be proportional to the electron density. Finally, if ionization by the maxwellian electrons occurs primarily from

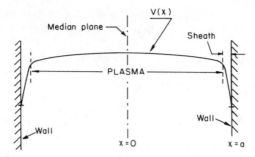

Figure 3.3. Anticipated potential variation in a plasma between plane parallel boundaries.

an excited state with the density of excited atoms proportional to the electron density n_e, the rate at which ions are produced would be expected to be proportional to n_e^2. We shall consider all three cases. The analysis follows that of Tonks and Langmuir (1929).

A. Planar Geometry

Although we shall discuss other configurations as well, let us examine in detail a plasma generated between two parallel plane electrodes at the same potential and a distance $2a$ apart. We anticipate a potential such as shown in Fig. 3.3, which falls very slowly through the plasma region where the electron and ion densities are approximately equal, with a sheath at the boundary of the order of the Debye shielding distance in which ion densities strongly predominate. It is assumed that fields perpendicular to the plane of Fig. 3.3 are zero.

The condition that the potential $V(x)$ must satisfy is Poisson's equation (3.13), with the electron density n_e given by the Boltzmann condition (3.12) and the ion density introduced as in Section 3.2. Suppose $g(x)$ is the ion generation rate, the number of ions per unit volume per unit time. Then the density at a point a distance x from the median plane is obtained by integrating the effect of all those ions formed between the median plane and the plane defined by x, taking into account the velocity which each group of ions acquires in falling from the potential at which it is formed to the potential $V(x)$. Thus

$$n_i(x) = \int_0^x \frac{g(x_1)}{v(x_1, x)} \, dx_1. \tag{3.29}$$

For the *collisionless* case, that is, for the case in which the ions, once formed, move to the walls with no collisions,* we can let

$$v = \sqrt{2e[V(x_1) - V(x)]/M} \tag{3.30}$$

*We shall examine in Chapter 4 the question of the reality of this assumption. For now let us assume that there are many plasmas for which this is a realistic assumption.

just as in Sec. 3.2. In so doing we have assumed that ions are formed with zero velocity. It is simplifying to use a dimensionless potential

$$\eta = -eV/kT,$$ (3.31)

where T is the electron temperature, leading to

$$v = \sqrt{2kT(\eta - \eta_1)/M}$$ (3.32)

where $\eta = \eta(x)$ and $\eta_1 = \eta(x_1)$. Poisson's equation, with η substituted for V, then becomes, after multiplying through by $\xi_0/n_0 e$,

$$\left(\frac{\xi_0 kT}{n_0 e^2}\right)\frac{d^2\eta}{dx^2} + \epsilon^{-\eta} - \frac{1}{n_0}\sqrt{\frac{M}{2kT}} \int_0^x \frac{g(x_1)}{\sqrt{\eta - \eta_1}} dx_1 = 0.$$ (3.33)

Take $\eta = 0$ at $x = 0$, so that n_0 is the electron density at the median plane. Now convert to a dimensionless length

$$\xi = x/L,$$ (3.34)

where L is a characteristic length to be chosen to simplify the problem. This yields, when we note that the coefficient of the first term in Eq. (3.33) is the square of the Debye shielding distance λ_D,

$$\left(\frac{\lambda_D}{L}\right)^2 \frac{d^2\eta}{d\xi^2} + \epsilon^{-\eta} - \frac{L}{n_0}\sqrt{\frac{M}{2kT}} \int_0^\xi \frac{g(\xi_1)}{\sqrt{\eta - \eta_1}} d\xi_1 = 0.$$ (3.35)

This equation can be expected to hold from wall to wall, that is, through the plasma and the sheath, and is termed the complete plasma–sheath equation.

Self (1963) obtains a first order differential equation by integrating Eq. (3.35). I prefer to arrive at the first order equation by a direct application of the momentum balance technique of Sec. 2.5. It is redundant to use both Poisson's equation and the momentum balance approach but, as we shall see, it will be useful to have both forms of the complete plasma–sheath equation. To apply the momentum balance equation consider the diode seen in Fig. 3.4. The left hand plate is taken to

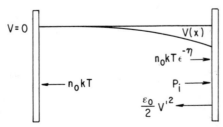

Figure 3.4. Imaginary diode used to derive the first order complete plasma–sheath equation.

be at the $x = 0$ plane of Fig. 3.3 and the right hand plate at the potential $V(x)$ corresponding to its position. If these planes are not to change, the plasma between them each must be emitting electrons at the same rate at which electrons reach the surface, and at the same temperature. The resulting electron pressure, $p = nkT$, is $n_0 kT$ on the left and $n_0 kT \exp (eV/kT)$ on the right. Electron pressure is the only pressure on the left, but the right hand surface as shown in Fig. 3.4 has, in addition, a pressure term p_i due to ion bombardment and an electric stress. Equating the sum of all pressure terms to zero yields

$$n_0 kT \left[1 - \exp \left(\frac{eV}{kT} \right) \right] - \int_0^x g(x_1) \sqrt{2eM[V(x_1) - V(x)]} \, dx_1 + \frac{\varepsilon_0 V'^2}{2} = 0.$$

Replacing V by η, as defined by Eq. (3.31), and x by ξ, as given by Eq. (3.34), and dividing through by $n_0 kT$, leads to

$$\frac{1}{2} \left(\frac{\lambda_D}{L} \right)^2 \eta'^2 + 1 - \epsilon^{-\eta} - \frac{L}{n_0} \sqrt{\frac{2M}{kT}} \int_0^\xi g(\xi_1) \sqrt{\eta - \eta_1} \, d\xi_1 = 0, \quad (3.36)$$

the desired first order differential equation. The equivalence of Eqs. (3.35) and (3.36) is demonstrated in the solution to Problem 3.4.

It is convenient to represent g by

$$g = \nu n_0 \epsilon^{-\gamma \eta} \quad (3.37)$$

where ν is a constant and γ assumes a value characterizing the ionization process. The case $\gamma = 0$ corresponds to uniform ionization $g = \nu n_0$, the situation which is likely to prevail if ionization is produced by primaries. The case $\gamma = 1$ is the situation in which the ionization is proportional to the electron density and, in this case, the constant ν is the number of ions per electron per second. The $\gamma = 2$ case represents ionization produced from an excited state which is, in turn, proportional to the electron density. If we now choose

$$L = \frac{1}{\nu} \sqrt{\frac{2kT}{M}}, \quad (3.38)$$

our complete plasma–sheath equation (3.35) becomes

$$\left(\frac{\lambda_D}{L} \right)^2 \frac{d^2 \eta}{d\xi^2} - \epsilon^{-\eta} - \int_0^\xi \frac{\exp (-\gamma \eta_1)}{\sqrt{\eta - \eta_1}} \, d\xi_1 = 0. \quad (3.39)$$

The relationship between L and the chamber width will be found but even before attempting a solution an approximate identification of L is possible. We can easily rewrite Eq. (3.38) as

$$Lvn_0 = n_0\sqrt{2kT/M}. \tag{3.40}$$

The Bohm criterion requires that v_i at the plasma boundary be approximately $\sqrt{kT/M}$. The density of ions at the boundary will be less than n_0 perhaps by a factor of about 2. The right hand side of Eq. (3.40) is therefore about three times the expected ion current density. The product of vn_0 is the rate of ion production in the center of the plasma. It appears that the length L will be approximately three times the plasma half width a.

For most laboratory plasmas 10^{-3} cm $< \lambda_D < 10^{-1}$ cm and if $L \approx 10$ cm, $(\lambda_D/L)^2$ lies between 10^{-4} and 10^{-8}. If $d^2\eta/d\xi^2$ is not very large compared to 1, the value of $\epsilon^{-\eta}$ at the plasma center, we can neglect the first term in Eq. (3.39) obtaining the plasma equation

$$\epsilon^\eta - \int_0^\xi \frac{\exp(-\gamma\eta_1)}{\sqrt{\eta - \eta_1}}\, d\xi_1 = 0. \tag{3.41}$$

The neglect of the term $d^2\eta/d\xi^2$ in Eq. (3.39) is equivalent to setting $n_e = n_i$ and is termed the plasma approximation. It would have simplified the analysis to have made this approximation directly, but the magnitude of the dropped term will be used to determine the range of validity of the plasma solution and the complete plasma–sheath equation will be necessary for joining the plasma to the sheath.

By appropriate changes of variables Harrison and Thompson (1959) transform Eq. (3.41) to a form for which they are able to use the Schlomilch transformation. In this manner they obtain closed form solutions for ξ as a function of η which are given here for the three cases of particular interest:

$\gamma = 0$, uniform ionization:

$$\xi = \frac{2}{\pi} e^{-\eta} D(\sqrt{\eta}) = \frac{2}{\pi} F(\sqrt{\eta}), \qquad \xi_0 = 0.3444 \tag{3.42}$$

$\gamma = 1$, ionization proportional to n_e:

$$\xi = \frac{2}{\pi}\left[D(\sqrt{\eta}) - \int_0^\eta D(\sqrt{s})\, ds \right], \qquad \xi_0 = 0.4046 \tag{3.43}$$

$\gamma = 2$, ionization proportional to n_e^2:

$$\xi = \frac{2}{\pi}\left[\sqrt{2}\, D(\sqrt{2\eta}) - \epsilon^\eta D(\sqrt{\eta}) \right], \qquad \xi_0 = 0.4920 \tag{3.44}$$

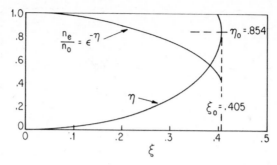

Figure 3.5. Solution of the plasma equation for the planar case with ionization rate proportional to the electron density n_e.

where

$$D(x) = \int_0^x \exp\left(t^2\right) dt \tag{3.45}$$

is related to the Dawson function* $F(x)$ defined by in Eq. (2.129).

Each of these functions has the general form displayed in Fig. 3.5 for the $\gamma = 1$ case. Each has η rising with ξ until ξ reaches a maximum value ξ_0 at $\eta = \eta_0 = 0.854$, and then doubles back. The portion of the curve corresponding to $\eta \geqslant 0.854$ has no physical significance. As $\xi \to \xi_0$ not only does $d\eta/d\xi \to \infty$ but so also does $d^2\eta/d\xi^2$, invalidating the neglect of the first term in Eq. (3.39). The exact solution must take the form shown in Fig. 3.6, departing from the plasma solution at the region of that curve where $d^2\eta/d\xi^2$ becomes so large that the neglect of the first term in Eq. (3.39) is no longer justified.

The place where this deviation from the plasma solution occurs can be judged from Fig. 3.7 where $d^2\eta/d\xi^2$ obtained from Eq. (3.43) by using

$$d^2\eta/d\xi^2 = -(d^2\xi/d\eta^2)/(d\xi/d\eta)^3 \tag{3.46}$$

is plotted as a function of ξ. For laboratory plasmas in which $10^{-8} < (\lambda_D/L)^2 < 10^{-4}$ the conditions appropriate to the plasma assumption [$n_e = n_i$, or neglect of the first term in Eq. (3.39)] hold to within a very small distance of the boundary. The point at which we can expect departure from the plasma solution is more easily obtained from Fig. 3.8, a plot of $d^2\eta/d\xi^2$ as a function of η. For example, if $(\lambda_D/L)^2 = 10^{-5}$, n_e and n_i would differ by 10% at a point which is virtually at $\xi = \xi_0$ as seen in Fig. 3.6, but is readily seen to occur at $\eta = 0.78$ in Fig. 3.7,

---

*There is a discrepancy in nomenclature which is apt to cause some confusion. Harrison and Thompson (1959) and Self (1963) refer to the function $D(x)$ as the Dawson function. Abramowitz and Stegun call $F(x) = \exp\left(-x^2\right) D(x)$ the Dawson function. The latter notation is used here.

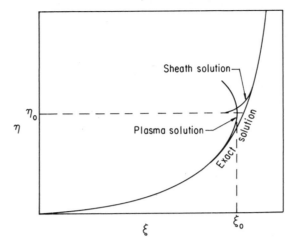

Figure 3.6. Various types of solution to the complete plasma–sheath equation.

substantially short of η_0. The merging of the plasma solution into the sheath, which occurs near this point, will be discussed in Sec. 3.8. The steepness of the plasmas solution at the point where this solution merges into the sheath guarantees that the sheath will be very small compared to the plasma width a. In that case we can identify $x/L = \xi$ at the plasma boundary with

$$\xi_0 = a/L \tag{3.47}$$

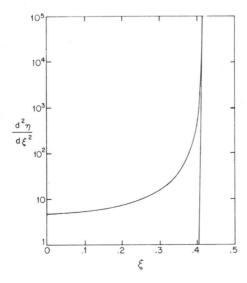

Figure 3.7. Laplacian of the potential as a function of distance from the center of the plasma in dimensionless units. The potential is the solution of the plasma equation for the planar case with ionization proportional to electron density.

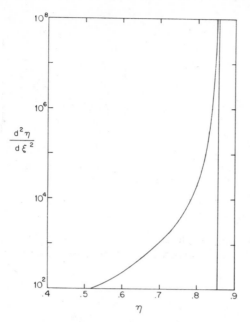

Figure 3.8. Laplacian of the potential as a function of potential for the case shown in Fig. 3.7.

yielding

$$L = a/0.3444 \qquad (3.48)$$

for uniform ionization,

$$L = a/0.4046 \qquad (3.49)$$

for ionization proportional to n_e, and

$$L = a/0.4920 \qquad (3.50)$$

for $g \sim n_e^2$.

The proper comparison among the three solutions is obtained by normalizing to ξ_0, that is, by plotting η as a function of $\xi/\xi_0 = x/a$ for each of the three solutions. This comparison is seen in Fig. 3.9.

The curves are sufficiently similar that it would be hard to distinguish among them experimentally and, as far as I am aware, no one has made sufficiently precise measurements of the plasma potential as a function of distance in a discharge to distinguish among the ionization modes, although Tonks and Langmuir (1929) and others have demonstrated that the general character of the potential variation is consistent with Fig. 3.9.

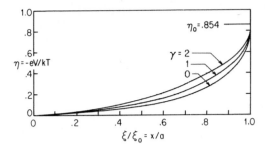

Figure 3.9. Potential variation between parallel planes for uniform ionization ($\gamma = 0$), ionization rate $\sim n_e (\gamma = 1)$, and ionization rate $\sim n_e^2 (\gamma = 2)$.

B. Cylindrical and Spherical Geometry

For a plasma generated inside cylindrical or spherical boundaries, the procedure parallels that for planar geometry but in these cases Eq. (3.29) would be replaced by

$$n_i = r^{-u} \int_0^r \frac{r_1^u g(r_1)}{v(r_1, r)} \, dr_1, \tag{3.51}$$

where $u = 1$ for the cylindrical case and $u = 2$ for the spherical case. ($u = 0$ yields the planar case, which I choose to handle separately.) As in the planar case, we let n_0 represent the electron or ion density at the center where the potential, represented by $\eta = -eV/kT$, is taken as zero. Then paralleling the steps which led to Eq. (3.35) we obtain

$$\left(\frac{\lambda_D}{L}\right)^2 \frac{1}{\xi^u} \frac{d}{d\xi} \left(\xi^u \frac{d\eta}{d\xi}\right) + \epsilon^{-\eta} - \frac{L}{n_0} \sqrt{\frac{M}{2kT}} \frac{1}{\xi^u} \int_0^\xi \frac{\xi_1^u g(\xi_1)}{\sqrt{\eta - \eta_1}} \, d\xi_1 = 0, \tag{3.52}$$

where ξ represents r/L. As in the planar case the solution in the plasma is obtained by dropping the first term.

Closed form solutions comparable to Eqs. (3.42)–(3.44) have not been found for these cases but series representations were developed by Tonks and Langmuir (1929). In general, the curves for the collisionless plasmas we have been considering have a form similar to the planar case but for cylindrical plasmas the turnaround potential is given by

$$\eta_0 = 1.155 \quad \text{at } \xi_0 = 0.7722 \tag{3.53}$$

when g is proportional to n_e, and

$$\eta_0 = 1.26 \quad \text{at } \xi_0 = 0.638 \tag{3.54}$$

for g constant.

For the spherical case, with g constant

$$\eta_0 = 1.50 \quad \text{at } \xi_0 = 0.818 \tag{3.55}$$

but the solution for the case $g \sim n_e$ has not been developed sufficiently to yield η_0 and ξ_0.

As far as I am aware solutions for the case $g \sim n_e^2$ have not been published for the cylindrical or spherical case.

3.6   Ion Velocity Distribution

The ion velocity distribution consistent with the assumptions made in Sec. 3.5, that is, that the ions are formed with zero velocity and make no collisions, can readily be calculated. I shall treat only the planar geometry.

Rewriting Eq. (3.29) without the integral yields

$$dn_i = [N(x_i)/v(x_1, x)] \, dx_1 \tag{3.56}$$

for the density of ions at x due to those formed between x_1 and $x_1 + dx_1$. Since the velocity is given by Eq. (3.30) we can transform the increment dx_1 to a velocity increment using

$$dv = \left[\frac{e/2M}{V(x_1) - V(x)} \right]^{1/2} V'(x_1) \, dx_1. \tag{3.57}$$

Since $V'(x)$ is negative we see that dv given by Eq. (3.56) is negative for a positive increment dx_1. Since we are interested in the spread in velocities comprising the group from dx_1, rather than the change in velocity when x_1 goes to $x_1 + dx_1$, we write

$$dv = \frac{e|V'(x_1)|}{Mv(x_1, x)} \, dx_1. \tag{3.58}$$

Solve this for dx_1 and insert the solution into Eq. (3.56), at the same time replacing g by the form given by Eq. (3.37). We obtain

$$dn_i = \frac{M}{e} \frac{vn_0 \exp(-\gamma\eta_1)}{|V'(x_1)|} \, dv. \tag{3.59}$$

With the dimensionless variables defined by Eqs. (3.31) and (3.34)

$$|V'(x_1)| = (kT/eL)(d\eta/d\xi)_{\xi_1},$$

and with a dimensionless velocity

$$u = v/\sqrt{2kT/M}, \tag{3.60}$$

we find

$$dn_i = 2n_0 \exp\left(-\gamma\eta_1\right)(d\xi/d\eta)_{\eta_1} du, \tag{3.61}$$

where $(d\xi/d\eta)$ evaluated at η_1 has replaced the reciprocal of $(d\eta/d\xi)$ evaluated at ξ_1. Our normalized distribution function is then

$$f_\eta(u) = (dn_i/du)/n_0 = 2 \exp\left(-\gamma\eta_1\right)(d\xi/d\eta)_{\eta_1}. \tag{3.62}$$

The derivative $(d\xi/d\eta)$ is readily evaluated for $\gamma = 0, 1$, or 2 using Eqs. (3.42)–(3.44). In each case the result is

$$\frac{d\xi}{d\eta} = \frac{2}{\pi} \epsilon^{\gamma\eta}\left[\frac{1}{2\sqrt{\eta}} - F(\sqrt{\eta})\right] \tag{3.63}$$

where $F(x) = \exp\left(-x^2\right)\int_0^x \exp\left(t^2\right) dt$ is the Dawson integral previously defined in Eq. (2.129). Surprisingly, perhaps, the resultant distribution function

$$f_\eta(u) = \frac{2}{\pi}\left[\frac{1}{\sqrt{\eta_1}} - 2F(\sqrt{\eta_1})\right] \tag{3.64}$$

is the same for $\gamma = 0, 1$, and 2. In fact, Chen (1962) shows that the distribution function is independent of the nature of the ionization function generally. Note that in our dimensionless units Eq. (3.30) becomes

$$\eta_1 = \eta - u^2, \tag{3.65}$$

so that Eq. (3.64) can be written in the more satisfactory form

$$f_\eta(u) = \frac{2}{\pi}\left[\frac{1}{\sqrt{\eta - u^2}} - 2F(\sqrt{\eta - u^2})\right]. \tag{3.66}$$

This is plotted in Fig. 3.10 for various values of η. The infinity at $u = \sqrt{\eta}$ occurs because $d\eta/d\xi = 0$ at $\eta = 0$. This causes no difficulty because the singularity is integrable. In reality, of course, the infinity would disappear if initial ion velocities or collisions were taken into account. It should not be expected that such changes will have a significant effect on the function $\eta(\xi)$.

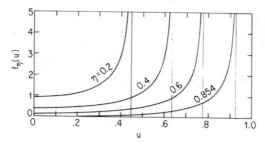

Figure 3.10. Ion velocity distribution in a collisionless plasma normalized to n_0, the electron density at the center of the plasma. The velocity u is in units of $\sqrt{2kT/M}$.

3.7 Ion Current Density

When normalized to n_0, the electron or ion density at the plasma center, we have found identical numbers of ions per unit velocity range for all ionization functions $g(x)$. This implies that the ion current from a planar plasma is exactly the same in all cases. Since this is so we can consider only the simple case of uniform ionization and obtain a result which is general. In this case the wall current density is simply given by

$$J = ega, \tag{3.67}$$

where g is the uniform ionization rate per unit volume and a is the distance over which ions are collected. But Eq. (3.47) gives

$$J = egL\,\xi_0 \tag{3.68}$$

and Eqs. (3.37) with $\gamma = 0$ and (3.38) give

$$J = 0.344\,n_0 e\sqrt{2kT/M} \tag{3.69}$$

using the value of $\xi_0 = 0.3444$ given by Eq. (3.42) for the uniform ionization case. I emphasize that Eq. (3.69), with the identical constant 0.344, holds independent of the value of γ which belongs in the ionization function.

It is interesting to compare this with the Bohm current given by Eq. (3.28). Since $0.344\sqrt{2}$ is close to 0.5 the values of J_i seem to differ by a factor of about 2. However, n_p, the plasma density at the boundary, is down from n_0 by a factor of about 2 so that the two formulas agree well.

For the cylindrical case Tonks and Langmuir (1929) found that the coefficient in Eq. (3.69) should be 0.319 and 0.270 for the cases g constant and $g \propto n_e$, respectively.

3.8   Ion Extraction Sheaths

The boundary of a plasma may face an electrode surface, an insulator, or an aperture from which ions are to be extracted. As this boundary is approached a transition occurs from the plasma region, in which ion and electron densities nearly balance, to the sheath region in which ion space charge predominates, as illustrated in Fig. 3.6. The complete plasma–sheath equation (3.35) does not appear subject to any simple analytic solution which bridges the transition between the two regions. We can for any particular potential on a boundary relative to the plasma easily approximate the sheath thickness and potential variation through the sheath by using the Child equation (2.8) or Eq. (2.47) with $V_0 = kT/2e$, obtained from the Bohm energy. Such approaches fail to take into account the penetration of electrons into the sheath and the spread in ion energies in the sheath. I therefore proceed to obtain more accurate solutions which can be used as a basis for determining the accuracy of such elementary solutions.

Before proceeding to do so let us note what sort of potentials may be expected across a sheath. The large size of the electron flux in a plasma compared to the ion flux will normally cause a plasma to go positive relative to the most positive electrode to which it has easy access, usually the anode of the discharge. If an electrode facing the plasma is at cathode potential, then the entire discharge voltage, usually 50–100 volts, plus the amount by which the plasma is positive relative to the anode, appears across the sheath. If an electrode is floating, or for an insulator surface, the potential across the sheath will be that value which will cause the ion and electron currents to exactly balance, that is, for which

$$n_0 e \exp\left(-eV_0/kT\right) \sqrt{kT/2\pi m} = 0.344 \, n_0 e \sqrt{2kT/M}, \qquad (3.70)$$

where n_0 is the electron density at the plasma center and V_0 the potential at the center relative to the wall. This equation presupposes that virtually all of the electrons are maxwellianized. If there is a substantial number of primaries, the floating potential of the electrode will approach that of the cathode. With negligible numerical error Eq. (3.70) leads to

$$V_0 = (kT/2e) \ln (2M/3m). \qquad (3.71)$$

The ratio of ion mass to electron mass can be replaced by $1836 \, M_A$ where M_A is the ion mass in atomic mass units (AMU). The predicted value of V_0 is plotted as a function of ion mass in Fig. 3.11. As a rough approximation, the plasma is seen to ride at a potential $5kT/e$ relative to a floating wall. Because electron current increases rapidly as this potential is decreased, the potential between a plasma and an anode is rarely much smaller.

Here, and elsewhere, we have assumed that the electrons, for the most part, are maxwellian and that their density therefore contains the Boltzmann factor $\epsilon^{-\eta}$. Even if equilibrium is established rapidly, we must expect that within the sheath the distribution in electron velocities directed away from the wall must be trun-

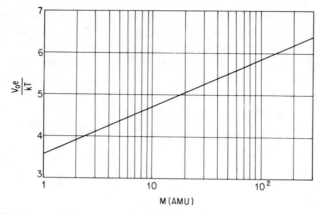

Figure 3.11. Plasma potential relative to a floating boundary as a function of ion mass, assuming maxwellian electrons at a temperature T.

cated. If the wall potential is sufficiently negative, the effect will be negligible, and in Fig. 3.11 we see that the boundary potentials must be at least $3kT/e$ below the potential at the edge of the plasma.

For purposes of estimation suppose the ion energy at the beam boundary to be the Bohm energy $kT/2$. The ion density at a point where the potential is lower by an amount ΔV is given by Eq. (3.23) to be

$$n_i = n_p\left(\frac{kT/2e}{kT/2e + \Delta V}\right)^{1/2} = n_p(1 + 2\Delta\eta)^{-1/2},$$

where n_p is the ion density at the plasma–sheath transition. For a maxwellian electron distribution the electron density would be

$$n_e = n_p\epsilon^{-\Delta\eta}.$$

For $\Delta\eta = 3$, which is about as small as we normally expect between the plasma boundary and a wall, $n_e/n_i \approx 0.13$. For this case the truncation of the electron distribution at the wall would produce about a 7% change in the charge density right at the wall and a lesser effect elsewhere in the sheath. The effect will be small. For values of $\eta_w > 4$ the effect rapidly becomes negligible.

A. The Sheath Approximation

For plasma densities used to generate high current density ion beams the plasma solution given by Eq. (3.42) or (3.44) is valid virtually up to the sheath transition. For these densities the sheath is so small compared to overall plasma dimensions that the fraction of the ion current generated in the sheath is negligible. The assumption that all ions are formed within a plasma in which the potential variation corresponds to the plasma solution enables us to find a generalized sheath solution. We shall call that assumption the sheath approximation. Because the velocity distribution of the ions at the plasma boundary, that is, at $\eta = \eta_0$, has been shown to be independent of the nature of the ionization function, we can simplify our treatment by assuming uniform ionization with no loss in generality. Subject to the assumption made, the potential variation through the sheath is independent of the ionization function, that is, independent of the value of γ in Eq. (3.37).

Take the form of the complete plasma–sheath equation given by Eq. (3.36) with g constant and $L = (n_0/g)\sqrt{2kT/M}$. We obtain

$$\frac{1}{2}\left(\frac{\lambda_D}{L}\right)^2 \eta'^2 + 1 - \epsilon^{-\eta} - 2\int_0^\xi \sqrt{\eta - \eta_1}\, d\xi_1 = 0. \tag{3.72}$$

To take this into the sheath region, using the sheath assumption, take $\xi > \xi_0$ and integrate only to ξ_0. Transform the integration variable from ξ_1 to η_1 using the plasma solution

$$\xi_1 = \frac{2}{\pi} F(\sqrt{\eta_1})$$

given by Eq. (3.42) for the uniform ionization case. As before,

$$F(x) = \exp(-x^2)\int_0^x \exp(t^2)\, dt. \tag{2.129}$$

Differentiating [or putting $\gamma = 0$ in Eq. (3.63)] yields

$$\frac{d\xi_1}{d\eta_1} = \frac{1}{\pi}\left[\frac{1}{\sqrt{\eta_1}} - 2F(\sqrt{\eta_1})\right],$$

enabling us to rewrite Eq. (3.72) as

$$\frac{1}{2}\left(\frac{\lambda_D}{L}\right)^2 \eta'^2 + 1 - \epsilon^{-\eta} - \frac{2}{\pi}\int_0^{\eta_0}\sqrt{\eta - \eta_1}\left[\frac{1}{\sqrt{\eta_1}} - 2F(\sqrt{\eta_1})\right]d\eta_1 = 0. \tag{3.73}$$

Now change the spatial variable $\xi = x/L$ to the variable $\zeta = (a - x)/\lambda_D$, where a is distance to the wall. This does two things. The length scale becomes the Debye length instead of the plasma dimension and distance is then measured from the wall into the sheath, instead of from the plasma center. The only term in Eq. (3.73) which is affected is the derivative

$$\eta' = (d\eta/d\xi) = -(L/\lambda_D)\, d\eta/d\zeta \tag{3.74}$$

and Eq. (3.73) becomes

$$\left(\frac{d\eta}{d\zeta}\right)^2 = \frac{4}{\pi} \int_0^{\eta_0} \sqrt{\eta - \eta_1} \left[\frac{1}{\sqrt{\eta_1}} - 2F(\sqrt{\eta_1}) \right] d\eta_1 - 2(1 - \epsilon^{-\eta}). \quad (3.75)$$

The first term in the integral is easily evaluated to yield

$$\int_0^{\eta_0} \left[\frac{(\eta - \eta_1)}{\eta_1} \right]^{1/2} d\eta_1 = \eta \left\{ \sin^{-1} \sqrt{\frac{\eta_0}{\eta}} + \left[\frac{(1 - \eta_0/\eta)\eta_0}{\eta} \right]^{1/2} \right\} \quad (3.76)$$

so that we find

$$\left(\frac{d\eta}{d\zeta}\right)^2 = f(\eta), \quad (3.77)$$

where

$$f(\eta) = \frac{4\eta}{\pi} \left[\sin^{-1} \sqrt{\frac{\eta_0}{\eta}} + \sqrt{\frac{\eta_0}{\eta} \left(1 - \frac{\eta_0}{\eta} \right)} \right]$$

$$- \frac{8}{\pi} \int_0^{\eta_0} \sqrt{\eta - \eta_1} \, F(\sqrt{\eta_1}) \, d\eta_1 - 2(1 - \epsilon^{-\eta}). \quad (3.78)$$

Since η decreases as ζ increases, Eq. (3.77) becomes

$$d\eta/d\zeta = -\sqrt{f(\eta)}, \quad (3.79)$$

leading to

$$\zeta(\eta) = \int_\eta^{\eta_w} \frac{d\eta_2}{\sqrt{f(\eta_2)}}, \quad (3.80)$$

where the subscript 2 has been used simply to distinguish between the integration variable and the variable η which appears as a limit. The quantity η_w represents the wall potential. To display $f(\eta)$ with all its complexities we rewrite Eq. (3.78) as

$$f(\eta) = \frac{4\eta}{\pi} \left[\sin^{-1} \sqrt{\frac{\eta_0}{\eta}} + \sqrt{\frac{\eta_0}{\eta} \left(1 - \frac{\eta_0}{\eta} \right)} \right]$$

$$- \frac{8}{\pi} \int_0^{\eta_0} \sqrt{\eta - \eta_1} \, \epsilon^{-\eta_1} \left(\int_0^{\sqrt{\eta_1}} \epsilon^{t^2} \, dt \right) d\eta_1 - 2 + 2\epsilon^{-\eta}. \quad (3.81)$$

**Table 3.1   Sheath Approximation Potential Variation
Near an $\eta = 20$ Wall**

η	ζ	η	ζ
20	0	3.3	11.402
19	0.496	3.0	11.838
18	1.003	2.8	12.157
17	1.520	2.6	12.503
16	2.049	2.4	12.883
15	2.591	2.2	13.309
14	3.149	2.0	13.797
13	3.723	1.9	14.071
12	4.316	1.8	14.372
11	4.931	1.7	14.706
10	5.573	1.6	15.082
9	6.245	1.5	15.514
8	6.956	1.4	16.022
7	7.715	1.4	16.643
6	8.539	1.2	17.439
5	9.456	1.1	18.550
4	10.516	1.0	20.362
3.6	11.003	0.9	25.009

It does not appear feasible to obtain an analytic solution but the results of a numerical integration for $n_w = 20$ are shown in Table 3.1 and Fig. 3.12. For any wall potential <20 the curve of Fig. 3.12 can simply be moved to the right.

Caruso and Cavaliere (1962) have treated the sheath problem starting with the complete plasma–sheath equation in the form given by Eq. (3.35). The mathematical complexities of their analysis tend to obscure the physical nature of their assumptions, but these appear to be identical to those made here, and Fig. 3.12 matches their curve for $\eta(\zeta)$.

B.   Exact Numerical Solution

We now have two solutions to the complete plasma–sheath equation, one valid in the plasma region and the other in the sheath region. These must join together in a manner such as illustrated in Fig. 3.6. Shelf (1963) carried out numerical integrations of Eq. (3.35) and, as a check, of Eq. (3.36), which he derived by integrating Eq. (3.35). His results are given in Table 3.2 for the three values of γ which we previously considered and various choices of a parameter $\alpha = \sqrt{2}(\lambda_D/L)$ and are plotted in Fig. 3.13 for the $\gamma = 1$ case.

To examine the solutions of Self in the sheath region convert the variable $\xi = x/L$ to

$$\zeta(\eta) = [\xi(6.5) - \xi(\eta)]\frac{L}{\lambda_D},$$

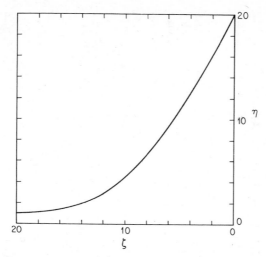

Figure 3.12. Sheath potential $\eta = -eV/kT$, where V is measured relative to the plasma center, as a function of ζ, distance from the wall in units of the Debye shielding distance, computed from Eqs. (3.80) and (3.81).

distance measured toward the plasma, in units of λ_D, with ζ taken as zero at $\eta = 6.5$. The curves obtained are seen in Fig. 3.14, and in Fig. 3.15 the curve for $\alpha = 10^{-3}$ is compared with the sheath solution of Fig. 3.12. The importance of this comparison has to do with the relative ease of applying Fig. 3.12, which has the advantage of generality, whereas each of Self's solutions is valid only for the particular values of γ and α for which it is carried out.

We see that for values of $\alpha < 10^{-3}$ the sheath potential very closely matches the asymptotic solution and we must ask what sort of plasmas correspond to this value of α. Suppose we imagine a laboratory plasma to have a half width $a \approx 0.1$ m, leading to a value of $L \approx 0.25$ m. For $\alpha = 10^{-3}$ we obtain, then, $\lambda_D \approx 1.77 \times 10^{-4}$ m. For a typical laboratory plasma with $kT/e = 5$ volts this corresponds to a plasma density of 8.9×10^{15} m$^{-3} = 8.9 \times 10^9$ cm^{-3}. This is a fairly low density laboratory plasma and down by a factor of about 100 from intense ion source plasma densities. It is clear that Fig. 3.12 provides a good basis for computing the potential variations in the sheath for most laboratory plasmas and virtually all ion source plasmas.

In estimating sheath thicknesses the Child equation often is used because of its simplicity. It is interesting to examine the validity of this approximation. Let V_0 be the potential at the plasma-to-sheath transition and apply Eqs. (2.8) and (3.69) to obtain

$$0.344\, n_0 e \sqrt{\frac{2kT}{M}} = \frac{4\varepsilon_0}{9} \sqrt{\frac{2e}{M}} \frac{(V_0 - V)^{3/2}}{(x_0 - x)^2},$$

Table 3.2 Plasma–Sheath Solutions of Self $[\alpha = \sqrt{2}(\lambda_D/L),\ \eta = -eV/kT,\ \text{and}\ \xi = x/L]$

| | $\gamma = 0$, Uniform Ionization | | | | $\gamma = 1$, Ionization $\sim n_e$ | | | | $\gamma = 2$, Ionization $\sim n_e^2$ | | | |
| | | ξ | | | | ξ | | | | ξ | | |
η	$\alpha = 0$	10^{-3}	$10^{-2.5}$	10^{-2}	0	10^{-3}	$10^{-2.5}$	10^{-2}	0	10^{-3}	$10^{-2.5}$	10^{-2}
0	0	0	0	0	0	0	0	0	0	0	0	0
0.1	0.1884	0.1884	0.1884	0.1885	0.1945	0.1945	0.1945	0.1946	0.2010	0.2010	0.2010	0.2011
0.2	0.2496	0.2496	0.2496	0.2498	0.2653	0.2653	0.2653	0.2655	0.2830	0.2830	0.2830	0.2831
0.3	0.2866	0.2867	0.2867	0.2871	0.3127	0.3127	0.3127	0.3130	0.3437	0.3437	0.3437	0.3439
0.4	0.3106	0.3107	0.3108	0.3116	0.3467	0.3467	0.3467	0.3472	0.3918	0.3918	0.3918	0.3920
0.5	0.3263	0.3263	0.3265	0.3283	0.3711	0.3711	0.3712	0.3722	0.4300	0.4300	0.4300	0.4305
0.6	0.3361	0.3362	0.3367	0.3399	0.3880	0.3881	0.3884	0.3904	0.4592	0.4593	0.4594	0.4604
0.7	0.3417	0.3419	0.3432	0.3483	0.3987	0.3988	0.3997	0.4036	0.4796	0.4796	0.4801	0.4827
0.8	0.3441	0.3451	0.3475	0.3546	0.4039	0.4046	0.4068	0.4133	0.4910	0.4910	0.4928	0.4986
0.9		0.3469	0.3504	0.3596		0.4074	0.4112	0.4206	0.4901	0.4956	0.4999	0.5098
1.0		0.3479	0.3525	0.3636		0.4088	0.4142	0.4263		0.4974	0.5040	0.5180
1.2		0.3492	0.3554	0.3698		0.4104	0.4180	0.4347		0.4992	0.5086	0.5290
1.4		0.3500	0.3575	0.3746		0.4112	0.4204	0.4407		0.5001	0.5114	0.5364
1.6		0.3506	0.3590	0.3784		0.4119	0.4222	0.4454		0.5008	0.5133	0.5418
1.8		0.3510	0.3603	0.3816		0.4123	0.4236	0.4493		0.5012	0.5148	0.5462
2.0		0.3514	0.3613	0.3844		0.4127	0.4247	0.4526		0.5016	0.5160	0.5499
2.5		0.3521	0.3635	0.3902		0.4135	0.4271	0.4593		0.5024	0.5184	0.5571
3.0		0.3527	0.3652	0.3949		0.4141	0.4289	0.4647		0.5030	0.5203	0.5628
3.5		0.3532	0.3666	0.3989		0.4146	0.4304	0.4692		0.5035	0.5218	0.5676
4.0		0.3536	0.3678	0.4025		0.4150	0.4317	0.4733		0.5039	0.5232	0.5718
4.5		0.3540	0.3690	0.4058		0.4154	0.4329	0.4769		0.5043	0.5244	0.5756
5.0		0.3543	0.3700	0.4088		0.4157	0.4340	0.4830		0.5047	0.5256	0.5791
5.5		0.3546	0.3710	0.4116		0.4161	0.4351	0.4834		0.5050	0.5266	0.5823
6.0		0.3549	0.3719	0.4143		0.4164	0.4360	0.4864		0.5053	0.5276	0.5854
6.5		0.3552	0.3728	0.4168		0.4167	0.4370	0.4892		0.5056	0.5285	0.5883

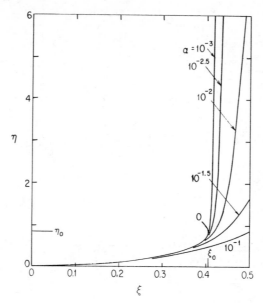

Figure 3.13. Exact plasma–sheath solutions for ionization proportional to $n_e(\gamma = 1)$. The parameter α is $\sqrt{2}\lambda_d/L$ and $\alpha = 0$ corresponds to the plasma approximation.

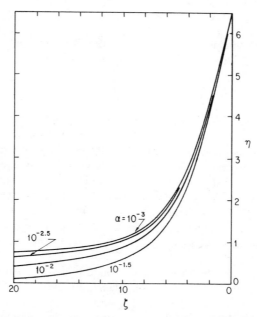

Figure 3.14. Potential variation for the $\gamma = 1$ case as a function of distance from a boundary in units of λ_D, from the numerical calculation of Self.

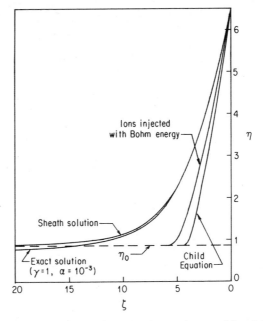

Figure 3.15. Comparison between various approximations to the potential variation in the sheath and the exact solution of Self for the case $\gamma = 1$ and $\alpha = 2\lambda_D/L = 10^{-3}$. Distance from the boundary, ξ, is in units of the Debye length λ_D.

where x_0 is sheath thickness and x is measured from the wall. Changing to $\eta = -eV/kT$ and $\xi = x/\lambda_D$ leads to

$$\zeta_0 - \zeta = 1.136 \left(\eta - \eta_0\right)^{3/4}. \qquad (3.82)$$

Take $\eta_0 = 0.854$ and obtain $\zeta_0 = 4.163$ by setting $\eta = 6.5$ at $\zeta = 0$. This solution, seen in Fig. 3.15, is clearly a poor approximation, and for η_w as low as 6.5 leads to a sheath thickness ζ_0, which is low by a factor of about 3 from the value that would be estimated from the Self solution or sheath approximation.

Another approach that is easy to carry out and is an improvement over the Child equation is to give the ions the Bohm energy $kT/2$ at the sheath boundary and apply Eq. (2.47). This procedure leads to a sheath thickness, as seen in Fig. 3.15, to be about half the correct sheath thickness for the case considered. The result of the sheath solution displayed in Fig. 3.12 is too easily applied to accept such poor approximations.

An exact value for a sheath thickness is hard to assign. Self suggests using the point where $\eta = \eta_0$ or $\xi = \xi_0$ as defining the sheath thickness. However, the slope of the $\eta(\zeta)$ curve is too small in this region to give the value so obtained much meaning. For example, in Fig. 3.14 it can be seen that the curves for $\alpha = 10^{-3}$ and $\alpha = 10^{-2.5}$ are nearly identical on the scale of the Debye length. However, if the $\eta = \eta_0$ criterion is used, the sheath thickness is quite different. For $\eta_w = 10$ Self derives a sheath thickness of $14.8\lambda_D$ for the $10^{-2.5}$ case and $17.6\lambda_D$ for the 10^{-3} case.

Because the Child equation is so frequently used to estimate sheath thickness it is useful to see how large η must be before this simple equation becomes an acceptable substitute for the sheath solution. From Eq. (3.82) we obtain, using the Child equation,

$$d\eta(d\zeta = (d\zeta/d\eta)^{-1} = -1.1737\eta^{1/4}\left[1 - (\eta_0/\eta)\right]^{1/4}. \qquad (3.83)$$

For the sheath solution this potential gradient is given by Eq. (3.79), so that we need to approximate $f(\eta)$ for large η. In Eq. (3.81) the term $\epsilon^{-\eta} \to 0$, the factor $\sqrt{\eta - \eta_1}$ in the integral becomes $\sqrt{\eta}\,(1 - \eta_1/2\eta)$, $\sin^{-1}\sqrt{\eta_0/\eta} \to \sqrt{\eta_0/\eta}$, and $\sqrt{1 - \eta_0/\eta} \to (1 - \eta_0/2\eta)$. With these approximations we obtain

$$f(\eta) = \frac{4\eta}{\pi}\left[\sqrt{\frac{\eta_0}{\eta}} + \sqrt{\frac{\eta_0}{\eta}}\left(1 - \frac{\eta_0}{2\eta}\right)\right] - \frac{8}{\pi}\sqrt{\eta}\,C_0 + \frac{4}{\pi}\frac{C_1}{\sqrt{\eta}} - 2, \qquad (3.84)$$

where

$$C_j = \int_0^{\eta_0} \eta_1^j\, F(\sqrt{\eta_1})\, d\eta_1. \qquad (3.85)$$

When $\sqrt{\eta}$ is factored out and terms are arranged in powers of η we obtain

$$f(\eta) = \sqrt{\eta}\left[\frac{8}{\pi}(\sqrt{\eta_0} - C_0) - \frac{2}{\sqrt{\eta}} - \frac{2}{\pi\eta}(\eta_0^{3/2} - 2C_1)\right]. \qquad (3.86)$$

As $\eta \to \infty$, $-\sqrt{f(\eta)}$ and the right side of Eq. (3.83) must become identical, relieving us of the necessity of evaluating C_0, except as a check. The third term in brackets can be neglected, at sufficiently large η, in comparison with the first two terms and we obtain

$$f(\eta) = \sqrt{\eta}\,(1.1737)^2\left[1 - (2/1.378\,\sqrt{\eta})\right]. \qquad (3.87)$$

Equation (3.79) then yields

$$(d\eta/d\zeta)_{\eta \gg 1} = -1.173\,\eta^{1/4}\left[1 - (1.452/\sqrt{\eta})\right]^{1/2}. \qquad (3.88)$$

A comparison shows that very large values of η are required before the potential gradient obtained from the Child equation and given by Eq. (3.83) matches the slope obtained from the sheath equation and given by (3.89) for large values of η. For example, at $\eta = 50$ the Child equation slope is 10% too high and even at η

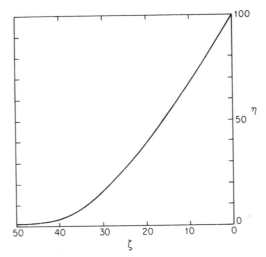

Figure 3.16. Sheath potential variation for $n_w = 100$. Curve for $\eta \leqslant 20$ is from Fig. 3.12 and for $\eta > 20$ from the large η approximation of Eq. (3.89).

$= 100$ it is 7% high. To achieve a slope that matches to within 1% would require a value of η of approximately 5000.

For values of $\eta > 20$, Eq. (3.88) provides a reasonably accurate approximation from which we obtain

$$\zeta = 0.852 \int_{\eta}^{\eta_w} (\sqrt{n_1} - 1.452)^{-1/2} \, d\eta_i .$$

This is readily integrated to yield

$$\zeta = 42.859 - 4.9485 \, (\eta - 1.452)^{1/2} - 1.1360 \, (\eta - 1.452)^{3/2} . \quad (3.89)$$

In Fig. 3.16 the sheath potential variation for a value of $\eta_w = 100$ is plotted using Eq. (3.89) for $\eta \leqslant 20$ and the data of Fig. 3.12 for $\eta \geqslant 20$. The match of the slopes at $\eta = 20$ confirms the validity of using Eq. (3.89) for $\eta > 20$.

3.9   The Cathode Double Sheath

If the negative potential surface facing the plasma is made to emit electrons, the potential variation and sheath thickness will be altered. The limiting case of a space charge limited cathode has been analyzed in detail by Crawford and Cannara (1964). If the sheath voltage is very large compared to kT_c/e, where T_c is the cathode temperature, we are justified in considering the electrons to be emitted with zero velocity. Thus on the cathode side this sheath resembles that shown in Fig. 2.14 and analyzed in Sec. 2.10. The plasma side, however, is very different, having all the complexities of Sec. 3.8: a distribution in ion velocities and with plasma electrons penetrating the sheath.

I limit my goal here to a calculation of the ratio of the electron to ion current in this case, that is, to investigate how Eq. (2.96) is altered when the ion emitting side is replaced by a plasma. Again, let us resort to the use of momentum balance technique of Sec. 2.5.

Instead of dealing with the exact case I simplify the problem by assuming that the ions all arrive at the sheath boundary with the Bohm energy $kT/2$ obtained from Eq. (3.27), and that the corresponding ion current density, from Eq. (3.28), is $n_p e \sqrt{kT/M}$, where n_p is the ion or electron density at the plasma-to-sheath transition. For purposes of this analysis the plasma surface is taken as a solid surface injecting directed ions and maxwellian electrons. The total pressure on this surface is then made up of three terms, the recoil from the ion injection, the pressure due to maxwellian electrons, and the rate at which momentum is delivered by electrons from the cathode:

$$F_p = (J_i/e) M\sqrt{kT/M} + n_p kT + (J_e/e) m\sqrt{2eV/m}.$$

But Eq. (3.28) gives $n_p = (J_i/e)\sqrt{M/kT}$, leading to

$$F_p = 2(J_i/e)\sqrt{MkT} + (J_e/e)\sqrt{2emV}. \qquad (3.90)$$

The force on the cathode is due only to the rate at which ions deliver momentum, leading to

$$F_c = (J_i/e)\sqrt{2eM(V + kT/2e)}. \qquad (3.91)$$

For $kT/2eV \ll 1$ we can write this as

$$F_c = J_i \sqrt{2MV/e}\,(1 + kT/4eV). \qquad (3.92)$$

Since F_p and F_c are in opposite directions we obtain

$$J_i(1 + kT/4eV) - J_i\sqrt{2kT/eV} - J_e\sqrt{m/M} = 0. \qquad (3.93)$$

If we drop $kT/4eV$ in comparison with one but retain $\sqrt{2kT/eV}$, we obtain

$$(J_e/J_i) = \sqrt{M/m}\,(1 - \sqrt{2kT/eV}). \qquad (3.94)$$

The difference between Eq. (3.94) and (2.96) can be substantial. For example, in a typical ion source plasma kT/eV might be 0.1, which would reduce J_e/J_i to 0.55 times the value of $\sqrt{M/m}$.

Although it is usually imagined that the discharges used in ion sources operate with space charge limited emission for which Eq. (3.94) would be appropriate, the cooling produced by the emitted electrons, a point which is discussed more fully in Sec. 7.4, makes this very unlikely. If the electron emission rises to the value for which the ion heating and electron cooling just balance, the electron emission would be at most one-half of the value given by Eq. (3.94).

3.10 Neutral Gas Density

A vital consideration for ion source design is the neutral gas density in the source plasma. Neutral gas escaping with the ions causes serious problems of many kinds. It is clear that in the laboratory handling the gas can be difficult and a major component of an ion source testing facility is the required pumping system. A more demanding requirement on the gas utilization efficiency (ratio of extracted ion per second to number of atoms per second fed to the source) results from the harmful effects of neutral gas in the ion acceleration region. These include electrode erosion, wide angle parasitics, electrode and source heating problems, and power loss. These will be discussed in detail in Chapter 5. For now let us concentrate on those factors influencing gas density and therefore gas utilization efficiency.

For $g \propto n_e$, we have, combining Eqs. (3.38) and (3.47),

$$a/0.4 = \left(1/\nu(\sqrt{2kT/M})\right), \tag{3.95}$$

where a is the half width of the plasma. The frequency ν (ions per electrode per second) is given by

$$\nu = n_a \overline{\sigma v}, \tag{3.96}$$

where n_a is the atomic density, σ is the cross section for ionization, v is the electron velocity, and $\overline{\sigma v}$ is the average value of the product, that is,

$$\overline{\sigma v} = \int_0^\infty \sigma v^3 \exp\left(\frac{-mv^2}{2kT}\right) dv \bigg/ \int_0^\infty v^2 \exp\left(\frac{-mv^2}{2kT}\right) dv. \tag{3.97}$$

The substitution of Eq. (3.96) into Eq. (3.95) yields

$$n_a = (0.4/a\overline{\sigma v})\sqrt{2kT/M}. \tag{3.98}$$

To illustrate the behavior of Eq. (3.98), $\overline{\sigma v}$ for cesium, mercury, and argon* are plotted in Fig. 3.17, assuming ionization from the ground state. The increase of $\overline{\sigma v}$ with temperature is so rapid compared with $\sqrt{T}$, in the range of electron temperatures at which ion sources operate, that increasing T lowers n_a in spite of the explicit dependence which appears in Eq. (3.93). This is illustrated in Fig. 3.18.

The calculation leading to the result seen in Fig. 3.18 is idealized in several ways, other than the assumption that the ionization is entirely by the maxwellian-

*Cesium data from Nygaard (1968) and Zapesochnyi and Aleksakhin (1969); mercury data from Nottingham (1939); argon data from Kieffer (1969) and Fletcher and Cowling (1973).

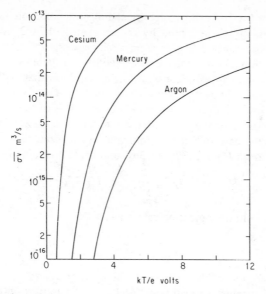

Figure 3.17. Mean value of σv as a function of electron temperature for cesium, mercury, and argon, where σ is the ionization of cross section for electrons of velocity v.

ized electrons. The assumption of an infinite parallel plane configuration is probably all right and it should be possible, in most cases, to make an estimate of an effective length a to use in Eq. (3.98). One idealization lies in treating the neutral gas as if the atoms were all in the ground state, that is, neglecting the fact that many will be in metastable states for which ionization cross sections will be much

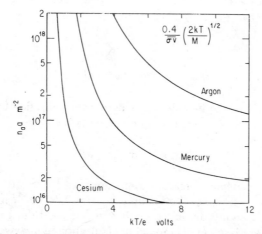

Figure 3.18. Product of neutral atom density and source thickness as a function of electron temperature for ionization produced only by interaction between maxwellian electrons and ground state atoms.

greater, especially for low energy electrons. For molecular gas inputs such as H_2 the situation is greatly complicated by the existence of both neutral atoms and molecules in the discharge and complicated chain reactions which can occur between the various ion and neutral species. Another idealization is in the treatment of the neutral gas density as a constant, an assumption that will break down at high plasma densities, as we shall see. Nevertheless, Eq. (3.93) is an important indication that a low value of n_a requires a large value of the source dimensions and, when the thermalized electrons do the ionizing, a large electron temperature.

Consider now the case of uniform ionization or ionization by primaries, the $\gamma = 0$ case, where from Eqs. (3.37), (3.38), and (3.48) we obtain

$$a/0.344 = (n_0/g) \sqrt{2kT/M}. \tag{3.99}$$

We can write the volume ionization rate g as

$$g = n_a \alpha n_0 \sigma_p v_p, \tag{3.100}$$

where α is the fraction of the electrons which are primaries at the center of the plasma, v_p is the velocity of primaries, and σ_p is the ionization cross section for electrons of velocity v_p. We obtain

$$n_a = (0.34/\alpha a \sigma_p v_p) \sqrt{2kT/M}. \tag{3.101}$$

The nonmaxwellianized electrons will of course have all velocities from the value $\sqrt{2eV/m}$ at which they are introduced down to those that make them part of the thermalized group. However, we shall see in Chapter 4 that the rate of loss of energy of the electrons can be expected to increase as energy decreases, so that once an electron loses energy its rate of decay increases. This will lead to a distribution that is peaked at the velocity at which the electrons are introduced. When we set $v = \sqrt{2eV/m}$ Eq. (3.96) becomes.

$$n_a = \frac{0.34}{\alpha a \sigma_p} \sqrt{\frac{kT}{eV} \frac{m}{M}}. \tag{3.102}$$

To decide which of the two situations dominate, ionization by primaries g_p or by thermalized electrons g_t, we note that at the plasma center

$$g_p/g_t = \alpha \sigma_p v_p / \overline{\sigma v}. \tag{3.103}$$

For mercury, for example, σ reaches a maximum of 5.3×10^{-20} m^2 at 30 volts where $v = 3.25 \times 10^6$ m/s. We obtain $\sigma_p v_p = 1.7 \times 10^{-13}$ m^3/s. At $kT/e = 4$ volts, where a mercury discharge might run, we see from Fig. 3.8 that $\overline{\sigma v}$ is about 10^{-14} m^3/s. Masek (1971) found the ionization by maxwellian and primaries to be approximately equal, corresponding to a value of $\alpha = 0.06$ in the particular mercury ion source plasmas studied by him. For argon $\sigma = 3 \times 10^{-20}$ m^2 at 80 volts, leading to $\sigma_p v_p = 1.6 \times 10^{-13}$ m^3/s. For $kT/e = 6$ volts, $\overline{\sigma v} \approx 4 \times 10^{-15}$ m^3/s, so that ionization by primaries dominates even if α is as low as 0.025. But higher energy electrons mix much more slowly and an argon plasma will have a greater fraction of primaries than mercury. For an argon plasma we expect virtually all of the ionization to be by primaries. For cesium exactly the reverse situation exists. The low energy electrons mix rapidly with the plasma leading to a very small value of α and this, coupled with the large value of $\overline{\sigma v}$, even at electron temperatures of 2 eV, leads to a plasma in which the ionization is almost entirely by the maxwellian electrons. The collision processes which produce the maxwellianization of the electrons are discussed in Chapter 4.

For the situation where ionization by maxwellian and ionization by primaries are both significant then we can show that

$$n_a = \frac{(0.37/a)}{(\alpha \sigma_p v_p)} \sqrt{2kT/M} \tag{3.104}$$

is an adequate approximation. The situation in which the ionization is proportional to n_e^2, to say nothing of those plasmas in which molecules undergo a complicated set of chain reactions leading to a variety of ions, is not discussed here.

To obtain the gas utilization efficiency let us use an equivalent current of neutral gas J_a obtained by multiplying the neutral gas efflux rate by the electronic charge e. Kinetic theory gives

$$J_a = n_a e \sqrt{kT_a/2\pi M}, \tag{3.105}$$

where T_a is the temperature of the neutral gas in the plasma region. For the case of ionization produced by thermalized electrons, Eq. (3.98) gives

$$J_a = (0.4e/a\overline{\sigma v})\,(k^2 T T_a/\pi M^2)^{1/2}.$$

When this is combined with Eq. (3.69) we find

$$J_i/J_a = 0.86 n_0 a\, \overline{\sigma v}\, \sqrt{2\pi M/kT_a}. \tag{3.106}$$

If the ionization is produced by primary electrons of energy eV, we obtain, using Eq. (3.101) instead of Eq. (3.98),

$$J_i/J_a = 2n_0 \alpha a \sigma_p\, \sqrt{\pi(M/m)\,(eV/kT_a)}. \tag{3.107}$$

For the case in which n_a is given by Eq. (3.104) we obtain

$$J_i/J_a = (0.93) n_0 a(\alpha \sigma_p v_p)\, \sqrt{2\pi M/kT_a}. \tag{3.108}$$

The mass utilization efficiency, the fraction of total throughput which emerges as ions, is given by

$$\eta = \frac{J_i/J_a}{1 + (J_i/J_a)}. \tag{3.109}$$

Equations (3.106) and (3.108) can lead to exceedingly high values of J_i/J_a. For example, consider a cesium plasma operating with an electron temperature of 2 eV where $\overline{\sigma v} = 2 \times 10^{-14}$ m^3/s and $n_0 = 5 \times 10^{17}$ m^{-3}. If we take $T_a = 0.05$ eV (580 K), Eq. (3.106) yields $J_i/J_a = 115$. It is clear that such large values of J_i/J_a would lead to a depletion of neutrals inconsistent with the assumption of uniform n_a implicit to the derivation of Eqs. (3.106) and (3.108). The equations should be regarded with suspicion when they lead to values greater than 1.

The effect of ionization on the depletion of the neutral density is discussed in Sec. 4.2, and the important role charge transfer plays in redistributing the atoms and in increasing neutral efflux is discussed in Sec. 4.3.

3.11   Characteristic Frequencies. Electromagnetic Wave Propagation

Although we shall have applications of waves in plasmas only in the section on RF ion sources, it would be remiss to have a chapter on plasmas with no mention of some of the characteristic frequencies that exist in plasmas and the consequences of these with respect to electromagnetic wave propagation. I shall make no attempt to derive expressions for these various frequencies or for the resonance and cutoff formulas which are presented. Unlike the rest of the material of this chapter such derivations can be found in any elementary text on plasma physics, for example, Chen (1974).

If, as a simple exercise, one imagined the ions of a plasma to be fixed in a uniform array and displaced some electrons from their uniform density distribution, then it can readily be shown that these electrons will oscillate about their equilibrium positions with the characteristic plasma angular frequency.

$$\omega_p = (ne^2/\varepsilon_0 m)^{1/2}. \tag{3.110}$$

If we ignore the thermal motion of the electrons this is a nonpropagating disturbance, but the inclusion of electron temperature effects leads to the Bohm–Gross dispersion relationship (not covered here), which describes the propagation of electron waves.

For the propagation of *electromagnetic* waves in an unmagnetized plasma the dispersion relationship is

$$\omega^2 - \omega_p^2 = \kappa^2 c^2 \tag{3.111}$$

where κ is the propagation constant which enters the term $\exp j(\kappa x - \omega t)$ representing a wave of angular frequency ω propagating in the $+x$ direction with a phase velocity ω/κ and group velocity $d\omega/d\kappa$. For the dispersion relation given by Eq. (3.111) κ is imaginary whenever $\omega < \omega_p$, implying a total reflection of radiation incident on a plasma. In that case (i.e., $\omega < \omega_p$)

$$\exp j(\kappa x - \omega t) = \exp(-x/\delta) \exp(-j\omega t), \qquad (3.112)$$

representing an evanescent oscillation whose amplitude in the medium decays exponentially with a characteristic distance.

$$\delta = c/(\omega_p^2 - \omega^2)^{1/2} \qquad (3.113)$$

for a typical ion source plasma $n = 5 \times 10^{17}$ m^{-3}, yielding a value of $\omega_p = 4 \times 10^{10}$ s^{-1} or a frequency $f_p = 6.3$ GHz. We see that microwave frequencies are required for the penetration of plasmas of densities of interest for intense ion sources. When $\omega < \omega_p$, except for values close to ω_p, the penetration depth δ will be of the order of 1 cm, so small compared to most plasmas of interest for intense ion sources that the radiation can be considered nonpenetrating.

When a plasma is in a magnetic field then, in addition to ω_p, there is a veritable zoo of characteristic frequencies. Electron oscillations along the field lines are unaffected by the field and still occur at the angular frequency ω_p. The combination of electrostatic and magnetic forces which act on electrons displaced perpendicular to the field, however, lead to an oscillation at a frequency

$$\omega_h = (\omega_p^2 + \omega_c^2)^{1/2}, \qquad (3.114)$$

called the upper hybrid frequency. The quantity

$$\omega_c = eB_0/m \qquad (3.115)$$

is the electron cyclotron frequency at the background magnetic field B_0.

To describe the propagation of electromagnetic waves through a magnetized plasma requires two other characteristic frequencies as well, ω_L and ω_R, defined by

$$\omega_{\genfrac{}{}{0pt}{}{L}{R}} = \tfrac{1}{2}\left[(\omega_c^2 + 4\omega_p^2)^{1/2} \mp \omega_c\right]. \qquad (3.116)$$

Very briefly, the behavior of electromagnetic waves in a plasma can be given in terms of the parameters ω_p, ω_c, ω_h, ω_L, and ω_R as follows.

Case 1:   Propagation $\perp B_0$, $E \| B_0$. There is no magnetic force associated with electron motion induced by the electric field of the wave. As could have been expected the dispersion relationship (3.111) still holds. The plasma reflects all ω < ω_p and is transparent (ignoring collisional effects) to all $\omega > \omega_p$.

Case 2: Propagation $\perp B_0$, $E \perp B_0$. No propagation (total reflection) $0 < \omega <$ ω_L and $\omega_h < \omega < \omega_R$. Resonance (energy absorption by electrons) at $\omega = \omega_h$.

Case 3: Propagation $\|B_0$, Left Hand (counterclockwise) Circular Polarization. No propagation $\omega < \omega_L$. No absorption $\omega > \omega_L$.

Case 4: Propagation $\|B$, Right Hand Polarization. *No propagation $\omega_c < \omega <$ ω_R. No absorption $\omega > \omega_R$. Resonance (energy absorption by electrons) at $\omega =$ ω_c.*

Problems

Section 3.2

3.1 A discharge tube has two planar electrodes inserted in the ends of a glass tube (Fig. 3.19). Suppose the electrodes are held at a potential difference of 5000 volts and the argon gas in the tube is illuminated by ionizing radiation (UV).

   (a) What is the maximum current read by the ammeter before a plasma will form, assuming no secondary ionization? (You may assume that charging of the walls will cause the particles to move parallel to the axis.)

   (b) What is the charge density in the tube? Assume that the intensity of the UV is such that the ionization probability for any atom is $0.1/s$.

   (c) What must the density of argon atoms be?

   (d) If the tube is at room temperature, what is the neutral gas pressure in torr?

   (e) Do you think the assumption that the electrons released do no ionization is a reasonable one in this case, knowing that the maximum cross section for ionization by electrons is about 3×10^{-16} cm^2?

Section 3.3

3.2 Show that Eq. (3.18) satisfies Poisson's equation for the same linearization approximation made in the derivation of Eq. (3.16).

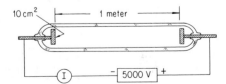

Figure 3.19. Discharge tube configuration for Problem 3.1.

3.3 Find the Debye shielding distance for the following plasmas: (a) carbon arc $n_e = 10^{18}$ cm^{-3}, $T = 3000$ K; (b) fusion plasma, $n_e = 10^{15}$ cm^{-3}, $T = 10^4$ eV; (c) ion source plasma, $n_e = 10^{12}$, $T = 5$ eV; (d) interplanetary space, $n_e = 10$ cm^{-3}, $T_e = 0.5$ eV; (e) interstellar space, $n_e = 1$ cm^{-3}, $T_e = 1$ eV.

Section 3.4

3.4 Suppose a bounded argon plasma has an electron temperature of 5 eV and the neutral gas occupying the same volume has a temperature of 580 K. What is the ratio of the average velocity components normal to the plasma surface of A$^+$ ions and A atoms? If the neutral atom density and the ion density at the edge of the plasma are equal, what is the ratio of the rates at which ions and atoms flow out of the plasma?

Section 3.5

3.5 Show that the differentiation of Eq. (3.36) yields the complete plasma–sheath equation (3.35). Note that

$$\frac{d}{dx} \int_0^x f(x, t)\, dt = f(x, x) + \int_0^x \frac{\partial}{\partial x} f(x, t)\, dt,$$

where $f(x, t)$ is any function of two variables for which the necessary derivative exists.

3.6 Find algebraic expressions equivalent to Eqs. (3.42)–(3.44) valid in the limit of small η. Compare the results with Fig. 3.9.

Section 3.6

3.7 Find the velocity distribution function for the ion current density. Normalize to $n_0 e \sqrt{2kT/M}$.

Section 3.7

3.8 Suppose that a xenon plasma has a density $n_0 = 5 \times 10^{11}$ cm^{-3} and an electron temperature of 6 eV. What ion current density is to be expected at the plasma boundary? (Assume planar geometry and singly charged ions.)

Section 3.8

3.9 An argon plasma faces a planar electrode with a large aperture and a fine grid (Fig. 3.20). For the parameters shown in Fig. 3.20 what potential must be applied to the fine grid so that the -30 volt equipotential is planar across the aperture?

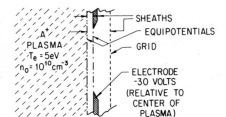

Figure 3.20. Geometry and various parameters for Problem 3.9.

Section 3.10

3.10   A mercury plasma operating with an electron temperature of 5 eV extracts ions from a length $a = 30$ cm. What is the mercury atom density to be expected in the plasma, if all atoms are in the ground state? Suppose the temperature of the neutral mercury vapor in the plasma to be 600 K and find the plasma density for which the mass utilization efficiency would be 90%. What ion current density would be obtained under these conditions?

3.11   An argon plasma with planar geometry has a thickness of 0.5 m, a plasma density of 2×10^{12} cm^{-3}, and an electron temperature of 6 eV. Five percent of the plasma electrons are 80 volt primaries. What is the gas utilization efficiency of this plasma as an ion source (fraction of argon ions which emerges from a hole in the boundary)? Take the neutral gas temperature to be 0.05 eV (580 K). The cross section for ionization of argon by 80 volt electrons is 3×10^{-16} cm^2.

Chapter 4

Collisional Effects

4.1 Types of Collisions and General Effects

Chapter 3 presented a view of plasmas in which collisions were totally neglected. Ionizations were assumed to occur but the depletion of neutrals due to this effect was ignored. As ions moved toward the boundary, interactions between ions and atoms and among the ions were neglected. We shall see that elastic collisions between ions and neutrals are in fact negligible but that charge transfer collisions, in which an ion and an atom interchange an electron, may be significant.

From the beginning we recognize that there are limitations on the effects that collisions can have. From the Bohm criterion, discussed in Sec. 3.4, we recognize that the plasma must contrive to deliver the ions to its boundary with an average energy of the order of $kT/2$. Interactions with neutral gas, which might slow the ions down, will have to be compensated for by stronger fields inside the plasma near the boundary. Interactions among the ions will not change their average drift velocity toward the boundary but will produce changes in the velocity distribution, that is, will tend to make the distribution maxwellian in the drifting reference frame.

4.2 Effect of Ionization on Neutral Gas Density

The development of the Tonks–Langmuir general theory of a plasma in Sec. 3.5 was done with reference to the closed system shown in Fig. 3.3. Ions striking the walls at $x = a$ and $x = -a$ would be reconverted to atoms and reenter the plasma. If the mean distance traveled by an atom before being ionized is greater than a, a uniform neutral atom density would be established consistent with an assumption implicit in the treatment as briefly discussed in the concluding remarks of Sec. 3.7. If this distance is small compared to a, then the gas density and ionization rate will be much greater near the wall than at the center.

Ion source configurations assume a wide variety of shapes and usually involve the substantial complication of magnetic fields as well but, for the present, it will be instructive to think of an ion source as being just a planar plasma, as illustrated in Fig. 3.3, but with the walls perforated for ion extraction. The efflux of ions and neutral atoms through these perforations must be compensated for by the addition of neutrals, say, in the median plane $x = 0$. In this case a mean distance to ionization $\ll a$ will lead to a peak in n_a at $x = 0$ and then a rise at the boundaries due to ions striking the closed fraction of the boundary.

It appears essential to examine this distance. If the ionization is produced by maxwellian electrons, the mean time that an atom can exist before being ionized is given by

$$\tau = 1/(n_e \overline{\sigma v}), \qquad (4.1)$$

where n_e is the electron density and $\overline{\sigma v}$ is the mean value of the product of ionization cross section and electron velocity plotted in Fig. 3.17. The distance an ion travels away from a plane in this time is given by

$$x_a = \sqrt{kT_a/M}/n_e \overline{\sigma v} \qquad (4.2)$$

where T_a is the temperature of the neutral gas. From Fig. 3.17 we obtain $\overline{\sigma v} = 8.4 \times 10^{-15}$ m^3/s for a 4 eV mercury plasma. If $kT_a \approx 0.05$ eV (580 K), we obtain

$$x_a \approx 1.8 \times 10^{16}/n_e$$

where x_a is in meters and n_e in electrons per cubic meter. If $n_e = 5 \times 10^{17}$ m^{-3}, a modest value for ion source plasmas, then $x_a = 3.7$ cm, a value inconsistent with the assumption of uniform n_a in a device whose dimensions are large compared to 3.7 cm.

The effective length for the decay of the neutral atom might be increased by charge transfer, which in effect speeds up the neutrals as ions are accelerated. This process is examined in the next section.

I am not aware of a treatment of a plasma which properly handles the variation in neutral atom density. It would be of interest to solve such a problem and I propose the following calculation. In Fig. 3.3 suppose the atoms are introduced at the median plane at the rate βJ_i and the planes $x = \pm a$ at a rate $(1 - \beta) J_i$ where β represents the fraction of the current to the boundary which is extracted to form an ion beam. Take into account the ionization depletion of neutral gas and charge transfer between atoms and ions to find $V(x)$, the ion current density and neutral atom efflux from a plasma. This problem will not be solvable analytically but should be subject to a computational approach.

4.3   Ion–Atom Collisions

If within the plasma the ions are streaming toward the boundaries with speeds much larger than the atom speeds, then the mean free path for a collision with a neutral atom is given by

$$\lambda = 1/(n_a \sigma_{ia}),\tag{4.3}$$

where n_a is the atom density and σ_{ia} the cross section for an ion–atom collision. The discussion of Sec. 4.2 notwithstanding, take n_a given by Eq. (3.67) and obtain

$$\frac{\lambda}{a} = \frac{\overline{\sigma v}}{0.4\sigma_{ia}} \sqrt{\frac{M}{2kT}}\tag{4.4}$$

for the case of a plasma in which ionization is by Maxwellian electrons. If the ionization is by primaries of energy eV, for which the ionization cross section is σ_p, Eq. (3.71) leads to

$$\frac{\lambda}{a} = \frac{\alpha \sigma_p}{0.34\sigma_{ia}} \sqrt{\frac{eV}{kT} \frac{M}{m}}.\tag{4.5}$$

Interestingly, λ/a turns out to be independent of a and independent of the plasma density.

Consider, for example, a cesium plasma with an electron temperature of 2 eV for which we can apply Eq. (4.4). Figure 3.8 yields $\overline{\sigma v} = 1.8 \times 10^{-14}$ m^3/s leading to

$$\lambda/a = 2.65 \times 10^{-17}/\sigma_{ia},\tag{4.6}$$

where σ_{ia} is in square meters. Elastic scattering cross sections may be about 10^{-19} m^2 leading to $\lambda/a = 265$, indicating that elastic collisions are of no importance. If we apply Eq. (4.5) to an argon plasma with an electron temperature of 5 eV, $\alpha = 0.08$, $V = 80$ volts, and $\sigma_p = 3 \times 10^{-20}$ m^2, we obtain

$$\lambda/a = 7.65 \times 10^{-18}/\sigma_{ia}.\tag{4.7}$$

Although this is smaller by a factor of 3.5, it still yields values of $\lambda/a \gg 1$ when $\sigma_{ia} \approx 10^{-19}$.

However, for very low energy a symmetric charge transfer process such as Cs^+ + Cs → Cs + Cs^+, can have a very large cross section. It is difficult to find accurate values at very low energy but from the work of Marino et al. (1962), Smith (1966), and Perel et al. (1965) it appears that the cesium charge transfer cross section might be as high as 5×10^{-14} cm^2 and that for argon the work of Bearman and Horne (1965) and of Smirnov and Chibisov (1965) indicate it might go as high as 10^{-14} cm^2. For the cesium plasma Eq. (4.6) then yields a value of $\lambda/a \approx 5$, and for the argon plasma Eq. (4.7) yields a value of $\lambda/a \approx 8$. These values are large enough that we may regard the ion flow as essentially unperturbed.

The effect on the neutral atom distribution and neutral efflux from the plasma, however, can be substantial. In Sec. 4.2 we saw that ionization can readily deplete the neutrals if neutrals move from the place of their introduction with thermal velocities. Interaction with the ions will accelerate neutrals away from their point of introduction. The charge transfer creation of fast neutrals heading toward the plasma boundary can also limit the obtained values of J_i/J_a. Equations (3.73) and (3.74) lead to values of 100 for easily obtainable plasma parameters, whereas values of J_i/J_a observed by Masek (1971) for mercury rise to about $8:1$ and those for cesium reported by Sohl et al. (1966) to about $13:1$, both under conditions where higher values should have been expected except for the effect of charge transfer. It appears likely that the production of neutrals which have energies large compared to room temperature thermal energy because of charge transfer is an important source of neutral escape from the plasma.

We started this section with the assumption that the ions are streaming past the neutrals with relatively high speeds. A glimpse at Fig. 3.9, however, shows a rather extensive region, of the order of halfway to the boundary from the center, over which the fields are very weak and ions move very slowly. The opportunity for mixing with the neutrals is optimal and it seems likely that the detailed study proposed at the end of Sec. 4.2 will show neutrals being carried along with ions to a larger measure than obtained in this section.

4.4   Collisions Between Charged Particles

To decide when ions once formed will move to the plasma boundary without collisions with other charged particles, as assumed in Sec. 3.5, or whether primary electrons will interact mainly with the plasma or with the neutral gas we need to understand the interactions between a charged particle and a plasma. As a step toward this goal let us examine the collision between two isolated charged particles.

As illustrated in Fig. 4.1 consider the interaction in a coordinate system which is fixed relative to the center of mass. The illustrated collision is for two particles of the same sign but the analysis would be identical for a collision between attracting, rather than repelling, particles. We illustrate the repulsive case only because the figure is slightly clearer when the paths do not intersect.

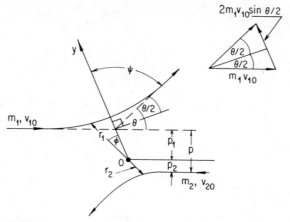

Figure 4.1. Repulsive coulomb collision between two particles in an inertial coordinate system which is stationary relative to the center of mass 0.

With the origin 0 at the center of mass we can let

$$m_1 r_1 = m_2 r_2 \tag{4.8}$$

and

$$m_1 v_{10} = m_2 v_{20}, \tag{4.9}$$

where v_{10} and v_{20} are the velocities of the two particles when they are a long distance apart and r_1 and r_2 describe the distances from 0 at any instant. Conservation of angular momentum requires that

$$m_1 v_{10} p_1 + m_2 v_{20} p_2 = (m_1 r_1^2 + m_2 r_2^2)\, d\phi/dt, \tag{4.10}$$

where p_1 and p_2 are the distances shown in Fig. 3.10 and ϕ is measured from the y axis, which is chosen to pass through the points of nearest approach. By substituting Eqs. (4.8) and (4.9) into Eq. (4.10) and letting $p_1 + p_2 = p$, we obtain

$$v_{10} p = r_1^2 \left(1 + \frac{m_1}{m_2}\right) \frac{d\phi}{dt}. \tag{4.11}$$

The distance p is called the collision parameter. From the diagram in the upper right of Fig. 4.1 showing the initial and final momentum of particle number 1 we see that its change in momentum is in the y direction and given by

$$\int_{-\infty}^{\infty} F_y\, dt = 2m_1 v_{10} \sin \frac{\theta}{2}, \tag{4.12}$$

where θ is the scattering angle. But $F_y = F \cos \phi$ where F is the coulomb force on the charged particle so that

$$\int_{-\infty}^{\infty} F_y \, dt = \frac{q_1 q_2}{4\pi\varepsilon_0} \int_{\phi_i}^{\phi_f} \frac{\cos \phi}{(r_1 + r_2)^2} \frac{d\phi}{d\phi/dt}, \tag{4.13}$$

where ϕ_i and ϕ_f are the initial and final values of ϕ. Using Eq. (4.8) we obtain

$$\int_{-\infty}^{\infty} F_y \, dt = \frac{q_1 q_2}{4\pi\varepsilon_0} \int_{\phi_i}^{\phi_f} \frac{\cos \phi \, d\phi}{r_1^2 (1 + m_1/m_2)^2 \, d\phi/dt}.$$

When Eq. (4.11) is applied this becomes

$$\int_{-\infty}^{\infty} F_y \, dt = \frac{q_1 q_2}{4\pi\varepsilon_0} \frac{1}{v_{10} p (1 + m_1/m_2)} \int_{\phi_i}^{\phi_f} \cos \phi \, d\phi.$$

Recognizing that $\phi_f = \psi = -\phi_i$ leads to

$$\int_{-\infty}^{\infty} F_y \, dt = \frac{2 q_1 q_2 \sin \psi}{4\pi\varepsilon_0 v_{10} p (1 + m_1/m_2)}. \tag{4.14}$$

But $\sin (\theta/2) = \cos \psi$ so that Eqs. (4.12) and (4.14) combine to yield

$$2 m_1 v_{10} \cos \psi = \frac{q_1 q_2 \sin \psi}{2\pi\varepsilon_0 v_{10} p (1 + m_1/m_2)},$$

which can be written as

$$\tan \psi = \frac{m_1 (m_1 + m_2)}{m_2} \frac{4\pi\varepsilon_0}{q_1 q_2} v_{10}^2 p. \tag{4.15}$$

We can easily show that Eq. (4.9) is equivalent to

$$v_{10} = v m_2/(m_1 + m_2),$$

where

$$v = v_{10} + v_{20} \tag{4.16}$$

is the relative velocity of the two particles, and when this is substituted for v_{10} in Eq. (4.15), the equation becomes

$$\tan \psi = 4\pi\varepsilon_0 \mu p v^2/q_1 q_2, \tag{4.17}$$

where μ is the reduced mass

$$\mu = m_1 m_2 / (m_1 + m_2).$$ (4.18)

For a $\theta = 90°$ collision, $\psi = 45°$ and Eq. (4.17) yields the scattering parameter

$$p_0 = q_1 q_2 / (4\pi\varepsilon_0 \mu v^2).$$ (4.19)

In terms of p_0, Eq. (4.17) becomes

$$\tan \psi = p / p_0.$$ (4.20)

For a collision between an electron and an ion the electron mass may be substituted for the reduced mass and the 90° scattering parameter becomes

$$p_0 = Ze^2 / 4\pi\varepsilon_0 m v^2,$$ (4.21)

where Ze is the charge on the ion. If the electron energy $mv^2/2$ is represented by eV_e, Eq. (4.21) becomes

$$p_0 = Z(e/\varepsilon_0)/(8\pi V_e).$$ (4.22)

The cross section for scattering through an angle of 90° or greater is

$$\sigma_0 = \pi p_0^2.$$

If the density of scatterers is n, then the mean free path between 90° scattering collisions is given by

$$\lambda_0 = (\sigma_0 n)^{-1} = (\pi p_0^2 n)^{-1}$$ (4.23)

and the mean time between collisions is

$$t_0 = \lambda_0 / v = (\pi v p_0^2 n)^{-1}.$$ (4.24)

Note that for θ very small ψ is close to $\pi/2$ and Eq. (4.20) indicates that $p \to \infty$ as $\psi \to \pi/2$. We shall see that the cumulative offset of distant encounters, each producing a small deflection in a random direction, outweighs the effect of close encounters. To show this a statistical treatment is necessary.

4.5 Scattering by a Plasma

When a stream of charged particles moves through a plasma, the effects of binary collisions between the incident particles and the ions and electrons in the plasma can be described by three parameters called diffusion coefficients. These three, which we write as $\langle \Delta v_\| \rangle$, $\langle (\Delta v_\|)^2 \rangle$, and $\langle (\Delta v_\perp)^2 \rangle$, represent, respectively, the mean change in velocity *per second* along the direction of motion, the mean square spread in velocity *per second* along the direction of motion, and the mean square velocity acquired perpendicular to the initial direction *per second*. These last two diffusion constants represent a diffusion in velocity space in exactly the same way that the constant D in the ordinary diffusion equation

$$\frac{\partial n}{\partial t} = D\nabla^2 n$$

represents diffusion in configuration space. The term $\langle \Delta v_\| \rangle$ was called by Chandrasekhar (1943) the coefficient of dynamical friction.

We shall not attempt to derive all of these diffusion constants but what is represented will be illustrated by carrying out one simple case, that of electrons scattered by ions. The main effect of each collision is a change in velocity, $\Delta v_\perp$, perpendicular to the initial velocity. Since each deflection is in a random direction we have the random walk problem translated into velocity space, so that the mean square value of $(\Delta v_\perp)$ after many collisions is the sum of the squares of $\Delta v_\perp$ acquired in each collision. If N is the number of collisions *per second*, then

$$\langle (\Delta v_\perp)^2 \rangle = \sum_{i=1}^{N} (\Delta v_\perp)_i^2 \tag{4.25}$$

represents the desired diffusion coefficient. We can see from Fig. 4.2a that

$$(\Delta v_\perp)^2 = v^2 \sin^2 \theta. \tag{4.26}$$

(a)

(b)

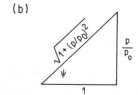

Figure 4.2. Relationships (*a*) between velocity increment and scattering angle θ and (*b*) between collision parameter and angle ψ.

But in Fig. 4.1 we see that $(\theta/2) = (\pi/2) - \psi$, so that Eq. (4.26) becomes

$$
\begin{aligned}
(\Delta v_\perp)^2 &= v^2 \sin^2 (\pi - 2\psi) \\
&= v^2 \sin^2 2\psi \\
&= 4v^2 \sin^2 \psi \cos^2 \psi.
\end{aligned}
$$

But Eq. (4.20), as illustrated in Fig. 4.2b, gives expressions for $\sin \psi$ and $\cos \psi$ which yield

$$
(\Delta v_\perp)^2 = 4v^2 (p/p_0)^2 \left[1 + (p/p_0)^2 \right]^{-2}. \tag{4.27}
$$

The number of plasma particles for which the collision parameter lies between p and $p + dp$ which are encountered per second is the number in the cylindrical shell of length v lying between the radii p and $p + dp$, that is, $2\pi n v p \, dp$ where n is the number of scatterers per unit volume. The summation indicated by Eq. (4.25) leads to

$$
\langle (\Delta v_\perp)^2 \rangle = 8\pi n v^3 \int_0^\infty p \left(\frac{p}{p_0} \right)^2 \left[1 + \left(\frac{p}{p_0} \right)^2 \right]^{-2} dp.
$$

Let $p/p_0 = s$ to obtain

$$
\langle (\Delta v_\perp)^2 \rangle = 8\pi n v^3 p_0^2 \int_0^\infty s^3 (1 + s^2)^{-2} \, ds. \tag{4.28}
$$

At large s the integrand approaches $1/s$. But the integral of $1/s$, from any lower limit to infinity, is infinity, a clearly unacceptable result. It would indicate that the interaction with distant particles is so strong that a particle in an extended plasma would instantaneously be deflected through $90°$, which means it could not move at all. Shielding efforts change this unrealistic conclusion.

In applying the results of Sec. 4.4 we implicitly assumed that each charged particle acts on each other charged particle through Coulomb's law and have neglected the shielding effect exhibited in Eq. (3.18). We can approximate the effect of the exponential falloff of potential around the moving charge by supposing that within a sphere of radius λ_D shielding effects can be neglected and that beyond the distance λ_D there is no interaction. We effect this by taking the upper limit in Eq. (4.28) to be

$$
\Lambda = \lambda_D / p_0. \tag{4.29}
$$

The 90° scattering parameter p_0 is of the order of atomic dimensions leading to a Λ which is usually in the range 10^4–10^5. For $\Lambda \gg 1$

$$\int_0^\Lambda s^2 (1 + s^2)^{-2} \, ds \approx \ln \Lambda - 0.5. \tag{4.30}$$

Usually Λ is so large that $\ln \Lambda$ is approximately 10 and we neglect the term 0.5 in Eq. (3.97) to obtain

$$\langle (\Delta v_\perp)^2 \rangle = 8\pi n v^3 p_0^2 \ln \Lambda. \tag{4.31}$$

Using p_0 obtained from Eq. (3.86) leads to

$$\langle (\Delta v_\perp)^2 \rangle = \frac{1}{2\pi} \left(\frac{e}{m} \right)^2 \left(\frac{e}{\varepsilon_0} \right)^2 \frac{nZ^2}{v} \ln \Lambda. \tag{4.32}$$

The procedure used gives an accurate result because the integral in Eq. (4.28) or Eq. (4.30) is insensitive to the upper limit, meaning that Eq. (4.32) is insensitive to the choice of Λ. For example, a typical value of $\ln \Lambda$ is 10 corresponding to $\Lambda = 22{,}000$. Suppose the effective cutoff should have been smaller by a factor of 2, leading to $\Lambda = 11{,}000$ instead of 22,000. This would change $\ln \Lambda$ to 9.3, a mere 7% difference. In almost all laboratory plasmas the use of $\ln \Lambda = 10$ is reasonably accurate and further calculation of $\ln \Lambda$ is usually avoided.

Following Spitzer (1962) we have derived only one of the diffusion coefficients and that one only for the special case of electron scattering by ions. It will be useful to have the results for other cases as well. These have been developed by Chandrasekhar (1942, 1943) and are presented below. In these equations let

$$m = \text{mass of the incident electron or ion}$$

$$m_f = \text{mass of the field or scattering particle}$$

$$Ze = \text{charge of the incident particle}$$

$$Z_f e = \text{charge of the scattering particle}$$

and

$$\alpha = v \sqrt{m_f / 2kT} \tag{4.33}$$

is the ratio of the velocity of the incident particle to the root mean square two dimensional velocity of the scatterers. Chandrasekhar's equations become

$$\langle \Delta v_\parallel \rangle = -A_D(m_f/2kT) \, (1 + m/m_f) \, G(\alpha), \tag{4.34}$$

$$\langle (\Delta v_\parallel)^2 \rangle = (A_D/v) \, G(\alpha), \tag{4.35}$$

and

$$\langle (\Delta v_\perp)^2 \rangle = (A_D/v)\left[\Phi(\alpha) - G(\alpha)\right], \tag{4.36}$$

where the "diffusion constant" A_D, in SI units, is given by

$$A_D = (1/2\pi)(e/m)^2 (e/\varepsilon_0)^2 nZ^2Z_f^2 \ln \Lambda. \tag{4.37}$$

In spite of redundancy, let me say that the density n of scatterers should be inserted in m^{-3}, (e/m) in C/kg and (e/ε_0) in V-m to yield A_D in m^3/s^4. The function Φ is the error function (erf) given by

$$\Phi(\alpha) = \frac{2}{\sqrt{\pi}} \int_0^\alpha \exp(-s^2)\, ds \tag{4.38}$$

and

$$G(\alpha) = \frac{\Phi(\alpha) - \alpha\Phi'(\alpha)}{2\alpha^2}. \tag{4.39}$$

But $\Phi'(\alpha) = (2/\sqrt{\pi}) \exp(-\alpha^2)$ and it is easy to see that for values of $\alpha > 2$

$$G(\alpha) \approx 0.5/\alpha^2 \tag{4.40}$$

is a good approximation to $G(\alpha)$. Since $\Phi \approx 1$ for values of $\alpha > 2$, $[\Phi(\alpha) - G(\alpha)] \approx 1$ for all $\alpha > 2$. In the case of electrons scattered by ions α is a very large number and we obtain from Eq. (4.36)

$$\langle (\Delta v_\perp)^2 \rangle = A_D/v,$$

which yields, as it must, Eq. (4.32), which was derived for that specific case.

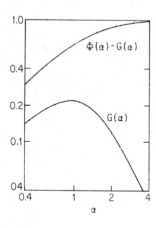

Figure 4.3. Functions $G(\alpha)$ and $\Phi(\alpha) - G(\alpha)$ required for evaluating diffusion coefficients.

For small α, say $\alpha < 0.5$, G can be represented by

$$G(\alpha) \approx 2\alpha/3\sqrt{\pi} \tag{4.41}$$

and

$$\Phi(\alpha) - G(\alpha) \approx 2G(\alpha). \tag{4.42}$$

Values of $G(\alpha)$ and $\Phi(\alpha) - G(\alpha)$ for values of α which are neither very small nor very large are shown in Fig. 4.3.

4.6 Relaxation Times

The diffusion coefficients can be used to introduce certain characteristic times called relaxation times. For example,

$$t_D = v^2/\langle (\Delta v_\perp)^2 \rangle \tag{4.43}$$

gives the time in which the particles deflect through approximately 90°, on the average, as a result of encounters with many plasma ions or electrons. Using Eq. (4.36) we obtain

$$t_D = \frac{v^3}{A_D[\Phi(\alpha) - G(\alpha)]}. \tag{4.44}$$

For the case of the scattering of electrons by ions we may use Eqs. (4.19) and (4.37) to obtain

$$t_D = \left[8\pi n v p_0^2 (\Phi - G)\ln\Lambda\right]^{-1}. \tag{4.45}$$

But for electrons scattering from ions $\alpha \gg 1$ and $(\Phi - G) = 1$, so that we can write

$$t_D = \left[8\pi n v p_0^2 \ln\Lambda\right]^{-1}. \tag{4.46}$$

If we compare this with the time for a 90° collision by a single close encounter as given by Eq. (4.24), we find that this time is less by a factor of $8\ln\Lambda$, approximately 80.

It is also of interest to define an energy exchange time

$$t_E = E^2/\langle (\Delta E)^2 \rangle. \tag{4.47}$$

To the first order

$$(\Delta E) = mv(\Delta v_\parallel), \tag{4.48}$$

so that

$$\langle (\Delta E)^2 \rangle = m^2 v^2 \langle (\Delta v_\parallel)^2 \rangle. \tag{4.49}$$

Equations (4.35) and (4.47) then yield

$$t_E = v^3 [4 A_D G(\alpha)]^{-1}. \tag{4.50}$$

The ratio of the deflection time t_D to the energy exchange time t_E is then

$$\frac{t_D}{t_E} = 4 \frac{G(\alpha)}{\Phi(\alpha) - G(\alpha)}, \tag{4.51}$$

which we see, using Eq. (4.42), approaches 2 for the case of a particle whose velocity is small compared to that of the scatterers. If $\alpha \gg 1$, that is, for fast particles,

$$t_D / t_E = 2/\alpha^2 \tag{4.52}$$

becoming $\ll 1$ as α gets large. However, for $\alpha \gg 1$ the neglect of terms due to deflection becomes invalid and Eq. (4.52) is not quite correct. Chandrasekhar (1941) has shown that for very large α, $G \ln \Lambda$ in Eq. (4.50) ($\ln \Lambda$ is contained in A_D) should be replaced by $0.5(1 + m_f/m)^{-2}$.

Problems

Section 4.4

4.1 A 100 volt electron moves through a cesium plasma of electron density 10^{12} cm^{-3}. What is the mean free path for a 90° scattering collision with a Cs$^+$ ion? What is the mean time between such collisions?

Section 4.5

4.2 A stream of electrons of velocity large compared to that of the thermalized electrons or ions moves through a plasma. Show that the rate of slowing down due to collisions with electrons is twice as great as for collisions with ions.

4.3 For the same electron stream show that $\langle (\Delta v_\perp)^2 \rangle$ is approximately the same for ions and electrons and that $\langle (\Delta v_\perp)^2 \rangle \gg \langle (\Delta v_\parallel)^2 \rangle$.

4.4 A 100 volt A$^+$ ion moves through an argon plasma of electron density 10^{12} cm^{-3}. Suppose the ion and electron velocity distributions correspond to a temperature of 5 eV. Find $\langle \Delta v_\parallel \rangle$ for collisions with ions and with electrons. Repeat for $(\Delta v_\parallel)^2$ and $(\Delta v_\perp)^2$.

Section 4.6

4.5 A 100 volt electron moves through a cesium plasma of electron density 10^{12} cm^{-3} and electron temperature 2.5 eV. Suppose the ion distribution corresponds to the same temperature as the electrons. What is the deflection time t_D? Compare with the result of Problem 4.1.

Chapter 5

Positive Ion Extraction and Acceleration

5.1 The Problem of Ion Extraction

When a plasma is generated ions drift toward the boundaries of the region as discussed in Sec. 3.5, reach the edge of the plasma with a directed kinetic energy of the order of kT, where T is the electron temperature in the plasma, and then fall through a sheath to the wall. The wall might be floating relative to the plasma, in which case the potential drop across the sheath would be of the order of $5kT/e$, it might be slightly less than the floating potential drop if the boundary were an anode drawing more electrons than ions, or it might be much greater. There would be no difficulty having a highly negative electrode face the plasma, if the effects of the bombardment by energetic ions could be tolerated. One convenient extraction system discussed in detail in Sec. 5.7 uses such a negative electrode in the form of a transparent fine mesh as a source of ions, but the use of this method is limited by heating and sputtering effects.

Normally the extraction of an ion beam requires that the wall have an aperture facing an accelerating electrode as shown in Fig. 5.1. With the correct electrode shapes (not those shown in Fig. 5.1) and the correct potentials it is possible to obtain a planar boundary to the plasma as seen in Fig. 1.1a. If, then, the potential difference between the two electrodes were reduced, the plasma would necessarily bulge out as shown in Fig. 5.1b. The shape of the surface of the plasma region is determined by the requirement that the solution of the unipolar space charge problem to the right of the plasma boundary match the near zero field at that surface considering the current density of slow ions delivered by the plasma.

Clearly the situation depicted in Fig. 5.1b is undesirable in that the accelerated ions diverge and many are intercepted by the accelerating electrode. By choosing

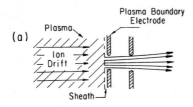

(a)

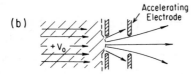

(b)

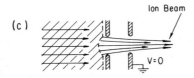

(c)

Figure 5.1. Ion extraction from a plasma (a) with a planar plasma boundary, (b) with inadequate potential difference between electrodes, and (c) with focusing produced by a concave plasma surface.

a potential difference larger than required for the situation depicted in Fig. 5.1a (or a smaller current density) one can achieve the plasma boundary configuration seen in Fig. 5.1c in which the ions are made to converge by the concavity of the plasma boundary. In this chapter we cover considerations relevant to the extraction of an ion beam including the conditions required to achieve the beam focusing illustrated in Fig. 5.1c.

5.2   Electrode Perveance and Poissance

Imagine a circular aperture of diameter a and an electrode spacing L as shown in Fig. 5.2 and let us make some approximations enabling us to arrive at order of magnitude relationships between current and voltage and electrode configuration that might be obscured if we immediately launched into precise calculations.

With $L \gg a$, as in Fig. 5.2a, the beam is clearly very far from satisfying the infinite plane conditions of Sec. 2.2 which led to the Child equation (2.8). Nevertheless, as we shall see in Sec. 5.3, with proper electrode shaping the ions can be made to travel in parallel paths and under that condition the relationships between potential, current density, and distance are exactly the same as for the infinite parallel plane geometry.

For voltage so large that the initial ion velocities can be neglected the current density given by Eq. (2.9a) is

$$J = \chi V^{3/2}/L^2,$$

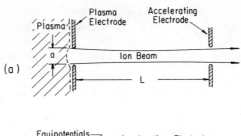

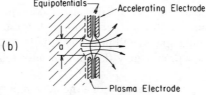

Figure 5.2. Ion extraction (a) with electrode separation $L \gg a$ and (b) with separation $L \ll a$.

where χ is defined by Eq. 2.9b and plotted in Fig. 2.2 as a function of ion mass. The total current,

$$I = \pi a^2 \chi V^{3/2} / 4L^2, \tag{5.1}$$

would appear to increase without limit as L is decreased. However, as L is decreased the bowing of equipotentials, as seen in Fig. 5.2b, limits the effective acceleration distance to a distance of the order of the diameter a. Setting $L = a$ gives then a measure of the maximum current

$$I_{\text{MAX}} \approx \chi V^{3/2}. \tag{5.2}$$

The crudeness of this calculation does not justify leaving the $\pi/4$ factor in the approximate equality (5.2). For slits of length $b \gg a$ we can replace this approximate equality by

$$I_{\text{MAX}} \approx \chi V^{3/2}(b/a). \tag{5.3}$$

The perveance, $P = I/V^{3/2}$, defined by Eq. (2.115) as an ion beam parameter, was originally used as a measure of electrode capability and is a function of the electrode geometry. We find that the maximum perveance obtainable for an electrode is given by

$$P_{\text{MAX}} \approx \chi b/a. \tag{5.4}$$

For ion guns, as accelerating systems are frequently termed, for which the perveance is close to the maximum, it can be expected, as shown in Fig. 5.2b, that the ions will be launched with a substantial angular spread although much of this might be compensated for by a highly concave emitter surface. In general it is helpful to stay well below the maximum perveance values. As for ion beams, the term poissance, the normalized perveance defined in Sec.2.9 and given by

$$\Pi = (I/V^{3/2})/(\chi b/a), \tag{2.116}$$

is useful.

We may say of a single aperture or slit that the maximum poissance is of the order of 1, and we may surmise that ion accelerating electrodes designed for producing beams of small angular spread will probably require poissance values small compared to 1. Values of approximately 0.15 are usually found consistent with small angular divergence.

Sometimes the perveance is compared with a theoretical perveance P_0 based on a parallel flow of a current density

$$J_0 = \chi V^{3/2}/L^2,$$

where V is the voltage between electrodes whose spacing is L. The corresponding current from an emitter of area S,

$$I_0 = \chi V^{3/2} S/L^2,$$

leads to an idealized perveance

$$P_0 = \chi S/L^2.$$

The ratio of actual perveance to P_0,

$$\frac{P}{P_0} = \frac{I}{V^{3/2}} \frac{L^2}{\chi S}, \tag{5.5}$$

is a measure of electrode performance compared to an ideal for the particular geometry, in comparison with the poissance Π, which is a measure of performance measured against an ultimate for any interelectrode spacing. Ther ratio of the two figures of merit,

$$\frac{\Pi}{P/P_0} = \frac{a}{b} \frac{S}{L^2}, \tag{5.6}$$

reduces to $(\pi r^2/L^2)$ for a circular emitter and to $(a/L)^2$ for a slit of width a.

5.3   Pierce Electrodes

Although ions must have a directed energy at a plasma boundary, the Bohm energy $kT/2$, usually of the order of 3 eV for ion source plasmas, the effect of this will be small if the ions are to be accelerated through several kilovolts or more. As a beginning, therefore, let us assume that the ions start from the plasma surface at rest. We also neglect the extremely small field that exists at the plasma surface, that is, at the plasma-to-sheath transition. Let us take as our first problem of electrode design that of designing electrodes to produce a parallel flow from a long slit. We shall see that the approach used here will be useful in designing electrodes for rectilinear flow generally, including cylindrically or spherically converging flow.

A.   Parallel Flow in Long Slit Beams

Recall the problem of space charge limited flow between infinite parallel plane electrodes, as covered in Sec. 2.2. Particles starting from rest at a surface where $dV/dx = 0$ move in straight lines toward a collector plane. The potential variation given by Eq. (2.7) can be written as

$$V(x) = (J/\chi)^{2/3} x^{4/3},$$

where $\chi = (4\varepsilon_0/9) \sqrt{2e/m}$. As in Sec. 2.2 let us carry out the analysis for electrons so that, with $V = 0$ at the emitter, V is nonnegative everywhere. Applying the results to ions involves no difficulties.

The appraoch of Pierce (1940, 1954) was to remove all of the beam on one side of a plane parallel to the direction of flow, say the $y = 0$ plane as shown in Fig. 5.3, and replace it with electrodes chosen to reproduce the boundary conditions so that the particles traveling in the region $y < 0$ have no way of sensing that they are not part of a flow between infinite parallel planes. By this means the space charge problem has been converted to the problem of finding a solution $V(x, y)$ of Laplace's equation in two dimensions,

$$\frac{\partial^2 V}{\partial x^2} + \frac{\partial^2 V}{\partial y^2} = 0,$$

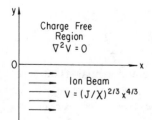

Figure 5.3.   Required conditions for producing parallel flow.

subject to the boundary conditions

$$\partial V/\partial y = 0 \quad \text{at } y = 0 \tag{5.7}$$

and

$$V(x, 0) = f(x), \tag{5.8}$$

where, for particles starting from rest at a zero field boundary,

$$f(x) = (J/\chi)^{2/3} x^{4/3}. \tag{5.9}$$

This is accomplished by taking the potential as the real part of

$$V + jW = f(x + jy), \tag{5.10}$$

where j is the imaginary unit $\sqrt{-1}$. It is evident that this satisfies Eq. (5.8). Furthermore, $(x + jy)^{4/3}$ is an analytic function.* For such functions the real and imaginary parts each satisfy Laplace's equation, so that requirement is met. Finally, since replacing j by $-j$ in Eq. (5.10) does not affect V, it is clear that replacing y by $-y$ will have no effect. As an analytic function which is even in y it is clear that $\partial V/\partial y = 0$ at $y = 0$. Thus

$$V(x, y) = (J/\chi)^{2/3} \left[\text{Real } (x + jy)\right]^{4/3} \tag{5.11}$$

is the potential that must be established in the region $y > 0$ in Fig. 5.3. Equation (5.11) may also be written as

$$V(x, y) = \left(\frac{J}{\chi}\right)^{2/3} (x^2 + y^2)^{2/3} \cos\left(\frac{4}{3} \tan^{-1} \frac{y}{x}\right).$$

The $V = 0$ equipotential requires that $(4/3) \tan^{-1} (y/x) = \pi/2$ so that the angle made with the x axis is $3\pi/8$ radians or $67.5°$ as seen in Fig. 5.4.

If we normalize to a potential $V_0 = V(x_0, 0)$ given by

$$V_0 = (J/\chi)^{2/3} x_0^{4/3},$$

then the normalized potential is given by

$$\frac{V}{V_0} = \left[\left(\frac{x}{x_0}\right)^2 + \left(\frac{y}{x_0}\right)^2\right]^{2/3} \cos\left(\frac{4}{3} \tan^{-1} \frac{y/x_0}{x/x_0}\right). \tag{5.12}$$

Except for the zero equipotential the shape of all positive equipotentials is the same

---

*An analytic function is one whose derivative exists and is independent of the direction in the complex plane in which the point is approached.

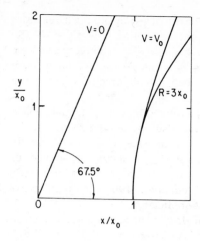

Figure 5.4. Equipotential surfaces for matching to a planar boundary of a space charge limited ion beam.

when normalized to V_0 and x_0. The equipotentials are shown in Fig. 5.4. All positive equipotentials (positive for electrons, negative for positive ions) are normal to the beam direction at the beam edge and at a distance x_0 from the beam edge are well matched by a cylinder of radius $3x_0$ as shown in Fig. 5.4. By placing electrodes with the potentials 0 and V_0 along the indicated surfaces we would achieve the desired effect of providing the boundary conditions which will cause the charged particles to propagate in straight lines even at the beam boundary.

To produce a long slit beam it would be necessary to put the same electrode structure on both sides of the beam as shown in Fig. 5.5. When the beam passes through the exit slit it is subject to the lens action of this slit leading to some divergence. The focal length of a long slit, to be found, for example, in Pierce (1954), is given by

$$f = \frac{2V_0}{V_2' - V_1'}, \tag{5.13}$$

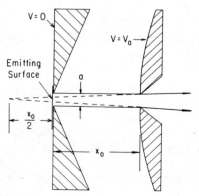

Figure 5.5. Electrodes for a long slit ion beam system showing divergence due to lens action at the exit aperture.

where V_2' and V_1' are the potential gradients on the downstream and upstream sides of the slit, respectively. In the case of space charge limited flow $V_1' = (4/3) V_0/x_0$ and if we take $V_2' = 0$, we obtain $f = -1.5x_0$, yielding the divergence seen in Fig. 5.5. This could, in principle, be avoided by a fine grid across the aperture in the V_0 electrode but heating and sputtering effects usually preclude doing so in intense ion beams.

In this treatment, we have, in the first place, assumed that the potentials inside the beam are parallel planes and then modified the result by using Eq. (5.13). For this treatment to be accurate we would require that the spacing x_0 be sufficiently large compared to the slit width a. As a minimum, x_0/a should be about 4 or larger for this treatment to be accurate. The effect of the aperture, termed the anode hole problem, is studied in greater detail by Kirstein et al. (1967), but the approximation given here should be adequate for most applications.

The emitting surface is of particular concern. If this surface is a solid electron emitter, or a surface ionization ion emitter, as discussed in Chapter 9, then all that is required for the considerations leading to Fig. 5.5 to be valid, other than $eV_0 \gg$ initial energy, is that the source provide, *as a minimum*, the space charge limited current density. When the emitting surface is the boundary of a plasma the relationship between the available current density and voltage, for a given x_0, must satisfy the Child equation. If the plasma supplies more than this current density, then the emitting surface will bulge outward, as in Fig. 5.1b, and the ion beam will be diverging in the interelectrode space.

B. Parallel Flow in Circular Beams

The approach is the same in the case of a beam of circular cross section. It is necessary to find electrode shapes that will cause the field outside the beam to match the potential given by Eq. (5.9) at the beam boundary and give zero normal field at the boundary, in this case $\partial V/\partial r = 0$ rather than $\partial V/\partial y = 0$. The simple analytic approach, which worked so well for a beam whose boundary was planar, cannot work in this case. Instead some other means must be found. Pierce (1940) describes an electrolytic tank approach* to find the desired electrode shapes presented in Fig. 5.6.

An accelerating system based on the curves of Fig. 5.6 is shown in Fig. 5.7. For a circular aperture the focal length, which may be found in Pierce (1954), is given by

$$f = \frac{4V_0}{V_2' - V_1'}. \tag{5.14}$$

With $V_2' = 0$ and $V_1' = (4/3) V_0/z_0$ we obtain $f = -3z_0$. The ions which approach the last aperture traveling parallel to the axis diverge as seen in Fig. 5.7.

*In 1940 there were no high speed digital computers with programs for finding solutions to Laplace's equation.

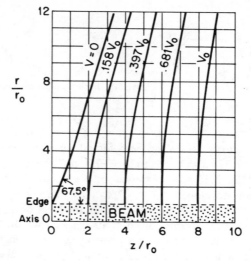

Figure 5.6. Equipotential surfaces which match to a space charge limited cylindrical beam.

In Figs. 5.5 and 5.7 the electrodes are carried far enough from the beam so that the fields at the beam are determined entirely by the electrode shapes. If the lateral dimensions are severely proscribed so that the electrodes cannot extend that far, then electrodes matching equipotentials for the desired field would need to be placed between the initial and final electrodes, as shown in Fig. 5.8.

C. Converging Flow

The electrodes shown in Figs. 5.5 and 5.7 produce beams of electrons or ions which travel in parallel paths between the emitter and accelerating electrode. It is usually preferable to have the beam converge in this region as illustrated in Fig.

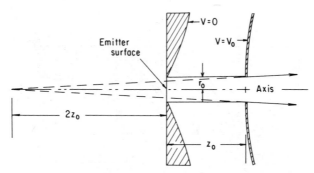

Figure 5.7. Electrodes for a cylindrical ion beam showing divergence due to lens action at the exit aperture.

Emitter

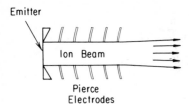

Ion Beam

Pierce
Electrodes

Figure 5.8. Beam forming electrodes under conditions of
limited lateral space.

5.1c. Since we have expressions for the potential variations in a cylindrically converging flow, Eq. (2.25) with β^2 given by $(-\beta)^2$ in Table 2.1, and in a spherically
converging flow, Eq. 2.26 with α^2 given by $(-\alpha)^2$ in the same table, we can
apply the techniques developed.

For example, Pierce (1940) presents a solution for a beam from a cylindrical
emitter made to converge to one-fourth of its original width. The equipotentials
outside the beam which reproduce this converging flow appear in Fig. 5.9. As
always, the zero potential surface makes an angle of 67.5° with respect to the
surface of the beam. Inside the beam, to the left of the exit slit, the equipotentials
are concentric cylinders except close to the exit aperture where divergent fields
will be encountered. In an actual accelerator electrodes matching these equipotentials must be placed on both sides of the converging beam.

The case of convergence toward a point along spherical radii is more complicated since there are two characteristic radii, the spherical radius r_0, shown in Fig.
5.10, and the emitter radius $a/2$ or the radius r_0 and the angle θ made by the
outermost converging ions and the axis. The electrodes shown in Fig. 5.10 are *not*
calculated shapes and we shall not attempt to develop Pierce electrode shapes for
any spherical convergent situation. What must be done, for any particular value
of r_0 and θ and a choice of the desired compression ratio within the electrode
system, is to first use Eq. (2.23) to find the current I, then with fixed I to obtain
$V(r)$ and then find the solution of Laplace's equation outside the beam which
matches the boundary conditions $V = V(r)$ and $\partial V/\partial\theta = 0$ at the beam edge.

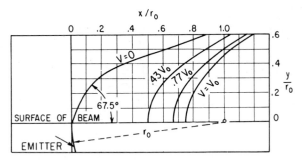

Figure 5.9. Equipotentials required for convergence toward an axis for the case of an exit slit one-
fourth the width of the emitter.

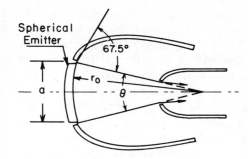

Figure 5.10. Schematic (not calculated) electrodes for a spherically convergent Pierce gun.

The calculation of perveance or poissance for any Pierce charged particle gun is readily carried out without recourse to the electrode shapes since the current voltage relationships are given by the equations developed in Sec. 2.2–2.4. In doing this it is necessary to be sure that the electrode separation is sufficiently large compared to the diameter of the exit aperture. A factor of 2 is probably adequate for accurate results. Examples of such calculations are given in Problems 5.1–5.3.

I have considered the defocusing action at the exit aperture but not the effect of space charge fields downstream of this electrode. It would not be difficult to do this utilizing the material covered in Secs. 2.11A and 2.11B, and Pierce (1954) covers both this and the effects of thermal velocities on the downstream beam. However, for ion beams this is unrealistic since the space charge spreading is largely canceled by the trapping of electrons inside the beam. This is discussed further in Sec. 5.5 and spreading produced by effects in the resultant beam plasma is covered in Chapter 6.

5.4 Emittance and Brightness

Broad ion beams, which are the primary focus of this book, are characterized typically by current density, angular divergence, and perveance. Perveance has been normalized in various ways, such as equivalent electron perveance, but I have chosen to invent a normalized perveance, which I call poissance. There are several other parameters by which ion beams are characterized. In particular, it is important to understand the meaning of the terms *emittance* and *brightness*, although each of these takes on its greatest importance for small ion beams such as those used for microscopy or microprocessing, or for beams that must deliver particles into a small area on a distant target.

The emittance, which is defined more carefully later, is roughly the product of beam diameter and spread in transverse momentum at each point. It is a quantity which one wants to minimize for any given current. The validity of treating this quantity as an invariant of an ion beam, independent of any electrostatic or magnetic fields through which the beam passes, relates to the Liouville theorem. To understand this theorem one imagines each particle to be represented, at each instant, by a point in the six dimensional phase space, that is, the coordinate system made of the spatial coordinates x, y, z and the momentum components p_x, p_y, and p_z. Imagine a large number of particles confined, at a given time, to a well defined volume of this space. Each particle will be moving along some path defined by such terms as $dx/dt = p_x/m$ and $d\mathbf{p}/dt = q(\mathbf{E} + \mathbf{p} \times \mathbf{B}/m)$. The constraint imposed by the Liouville theorem is that the volume in phase space must remain constant in any stationary combination of electric and magnetic fields.

For a group of ions, all accelerated through the same potential difference, and traveling with small angles relative to, say, the z direction, we might surmise that they have a very small spread in z directed velocities and that this spread is an invariant. Then the four dimensional volume occupied by the particles with coordinates x, y, p_x, p_y should be an invariant of the beam. For a beam with cylindrical symmetry this implies that the two dimensional area with coordinates r and p_r should be an invariant. For ions which make a small angle α relative to the z axis, $p_r = \alpha p_z$ where p_z can usually be taken as a constant. A typical envelope of the points representing the p_r and r values of the ions at a given cross section might have a shape such as that shown in Fig. 5.11. We define the transverse emittance, usually referred to simply as emittance, as the enclosed area in the $p_r \cdot r$ plane. Since the particle density as a function of radius as well as the angular distribution at a given point are likely to be gaussian-like, it is sometimes necessary to specify that the emittance diagram includes a given fraction such as 80% of the ions.

Since p_z is a constant over a cross section it is common to see emittance plots in which α alone is used as the ordinate. In that case the area will change when the beam undergoes acceleration.

Since the arguments (somewhat flawed, as pointed out below) lead to a constant transverse emittance it is, in principle, possible to transform the beam whose emittance diagram is shown in Fig. 5.11 to a long narrow vertical ellipse near $r = 0$, representing a small spot with large angular divergence, or a horizontal ellipse enclosing only points near $\alpha = 0$ representing a large diameter highly parallel beam. Even though the emittance may be regarded as a constant, poor optics might produce a beam whose emittance diagram, for example, resembled that in Fig.

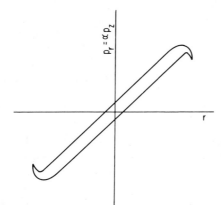

Figure 5.11. Possible emittance diagram for a cylindrically symmetric ion beam.

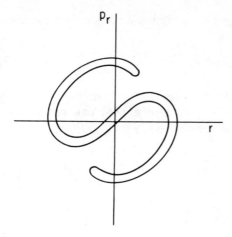

Figure 5.12. Emittance diagram of an ion beam with "air" folded in.

5.12. Such an emittance diagram is said to have folded in air and it would be difficult to transform that to a compact area in an α–r diagram. Its effective emittance has been made larger than its real emittance by poor ion optics.

The transformation of the six dimensional invariant phase space volume to an invariant two dimensional emittance by considering v_z to be constant is useful but not rigorous. It is possible to reduce the transverse emittance by increasing the spread in z directed velocities: Whealton (1984), for example, described an accelerating system in which the calculated trajectories decrease in overall diameter by a factor of 10 while the spread in random transverse velocities increases by only a factor of 3, corresponding to a decrease in transverse emittance by a factor of about 3.

A beam of light is characterized by an invariant power per unit area per unit solid angle. The corresponding quantity for an ion beam,

$$B = J/\Omega, \tag{5.15}$$

where J is the current per unit area and Ω the spread in solid angle at each point, is called the brightness. The solid angular spread is proportional to α^2, where α is the angular spread at each point, and the area is proportional to r^2 so that, neglecting constant factors, Eq. (5.15) becomes

$$B = I/r^2\alpha^2. \tag{5.16}$$

But the emittance may be approximated by

$$\varepsilon = r^2\alpha^2 p_z^2 \tag{5.17}$$

and since p_z^2 is proportional to the voltage V through which the ions were accelerated we can write

$$B = IV/\mathcal{E}^2. \tag{5.18}$$

D. B. Langmuir (1937) showed that the current per unit area, for particles originating with a maxwellian distribution, as electrons from a thermionic emitter, is limited in the current density which can be achieved at any point to

$$J = J_0 \left(\frac{eV}{kT} + 1 \right) \sin^2 \alpha, \tag{5.19}$$

which can usually be approximated by

$$J = J_0 \alpha^2 eV/kT. \tag{5.20}$$

This would give a maximum brightness which may be approximated by

$$B = J_0 eV/kT. \tag{5.21}$$

While the ions do not originate with a maxwellian velocity, as seen in Fig. 3.10, for example, we can, as a reasonable approximation, take T in Eq. (5.21) as the electron temperature in the source plasma to obtain an upper limit to the achievable brightness.

A more extensive analysis of emittance and brightness and other parameters by which ion beams may be characterized has been given by Schumacher and Fink (1973; also Fink and Schumacher, 1974, 1975).

5.5 Multiaperture Accel–Decel Extraction

We saw from the discussion of Sec. 5.1 leading to the approximate equality (5.4) that perveance of a single aperture is limited to $\chi = (4\varepsilon_0/9) \sqrt{2e/M}$ or to $\chi b/a$ for a long slit. This limitation, equivalent to a maximum poissance of 1, is emphasized in the calculations of poissance in the examples given in Problems 5.3–5.5 To obtain larger currents it is necessary to use an array of apertures or slits such as shown in Fig. 1.1. For intense ion sources the number of circular apertures may run to many hundreds even thousands of apertures, as indicated by the solution to Problem 5.6.

To achieve a high transparency of the plasma electrode requires close packing of the apertures or slits, which eliminates the lateral space required for a pure Pierce geometry. Furthermore, it is essential that ion accelerating systems for high current sources first accelerate the beam to a higher than desired energy and then decelerate the ions, hence the name accel–decel. The reason for this relates to the necessity of preventing electron backflow to the ion source.

This point requires elaboration. The ion current flowing between the plasma electrode and the accel electrode, using the nomenclature of Fig. 1.1, is essentially a space charge limited flow corresponding to the electrode spacing, albeit somewhat diluted by the spaces between apertures and the recessed plasma surfaces. At the exit plane this beam can propagate only a distance of the order of the electrode spacing before the potential could be expected to rise to the source potential and reflect the beam, a process known as beam stalling and discussed previously in Sec. 2.11C.

That this normally does not occur is due to neutralization of the positive ion space charge by trapped electrons. For the beam to stall, the potential must rise to a maximum. An ion beam necessarily creates electrons by secondary emission at all surfaces struck by ions and, for sufficiently energetic ions, by ionization of the ambient gas. These electrons become trapped in the potential well created by the positive ion space charge and interact with each other and with the ions to create a plasma which we shall call the beam plasma. This plasma is comprised not only of the fast beam ions and thermalized electrons but may contain slow ions created by charge transfer between the beam ions and the neutral gas. Even when the fraction of the beam which undergoes charge transfer is small the ions created move out of the region occupied by the beam so slowly compared to the beam ions that their density may be much larger than that of the beam ions. The electron density must approximately equal the sum of the ion densities. Much more will be said about this plasma in Chapter 6.

Imagine now, again with reference to Fig. 1.1, that the potential were to fall monotonically between the source plasma and the beam plasma. We would have a double sheath configuration with ions flowing from the source plasma to the beam plasma and electrons flowing in the reverse direction. As shown in Sec. 2.10, this would require an electron current greater than the ion current by a factor of $\sqrt{M/m}$. Such an electron current, if it could be supplied, would result in delivering so much power to the source that the source would quickly convert to a molten glob. Failure to supply the electrons would result in depletion of electrons from the beam and consequent stalling of the beam.

It is essential therefore that potential along the beam assume the form shown in Fig. 5.13, resulting in an acceleration and then deceleration of the ions to their final velocity. Electrons from the beam plasma then cannot get back to the source. Slow ions produced by charge transfer in the beam can be accelerated by the decel voltage back to the accel electrode and, in fact, sputtering of this electrode by charge transfer ions is often the major source of ion source lifetime limitation.

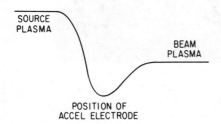

Figure 5.13. Requisite potential variation between the source plasma and beam plasma.

Insofar as the foregoing discussion suggests that the three electrode system shown in Fig. 1.1 *must* always be used to produce large currents, it is misleading. In fact Problem 2.10 illustrates an accel–decel system with only a single electrode. Such fine mesh accelerating systems will be discussed in detail in Sec. 5.7. Normally at least two electrodes are needed, with the plasma electrode serving the purpose of avoiding direct bombardment of the accel electrode by high energy ions. The beam itself, before it stalls, will rise to whatever potential is necessary to trap the electrons that are required for neutralization. Imagine Fig. 1.1 without the decel electrode. The beam would then rise to a potential sufficiently positive relative to the surroundings to trap the space charge neutralizing electrons. Even if the decel power supply voltage were reduced to zero the potential along a beamlet would show the accel–decel character of Fig. 5.13, although the deceleration would be minimal. In general the beam will have to go positive relative to the most positive potential in its surroundings. It is important therefore that the surroundings do not include potentials more positive than the desired beam potential.

In one run with an ion source some electrodes at the source potential were inadvertently exposed to the beam. Only the feeblest of currents could be obtained and the trajectories of the unneutralized beamlets are shown in Fig. 5.14. The positive electrodes are seen as the dark blob in the lower left and the bright object in the upper center. The space charge spreading of the unneutralized beam is easily observed and matches the calculated spreading shown in Fig. 2.16.

Since it is unnecessary to have more than two multiapertured electrodes to accelerate a beam, avoid bombardment of the accel electrode, and deter backstreaming of electrons from the beam plasma the significant question becomes "Why

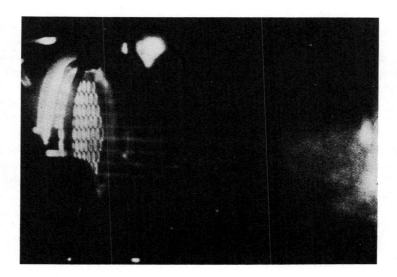

Figure 5.14. Photograph of space charge spreading in an unneutralized beam produced by exposing source potential electrodes to the ion beam (Reproduced from Ernstene et al., 1961).

have three electrodes?'' One answer relates to the greater currents that can be drawn from the source at the larger gap voltages. That is, for a desired beam voltage, say, 10 kV, and for given aperture sizes and spacing between the plasma and accel electrodes more current per aperture can be handled by accelerating to 20 kV and then decelerating to 10 kV than simply by accelerating to 10 kV. It would be possible to do as well at 10 kV without a large accel–beam voltage ratio by scaling down the geometry but when hole sizes become too small electrode alignment becomes a major problem. This point may be simply stated as follows: Large accel voltages make possible coarser geometry. As we shall see, however, large accel voltages have serious disadvantages as well.

The compelling reason is often a different one and relates to the shape which the beam plasma surface assumes. At the boundary between the beam plasma and the decelerating field the potential gradient must go to zero, as shown in Fig. 5.13. Because the positive space charge causes the unneutralized beam to be positive in the center relative to the outside, the beam plasma surface will be convex toward the accel electrode, as in Fig. 5.15a. The resultant equipotentials shown there

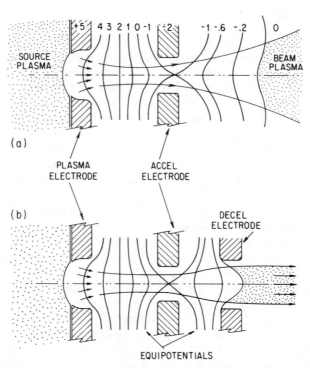

Figure 5.15. Equipotentials and ion trajectories for an accel–decel system (a) in which there is no decel electrode and (b) in which a decel electrode is used. The equipotentials and trajectories are not calculated curves but are drawn to be illustrative of the nature of the shapes to be expected. Numbers on the equipotentials are kilovolts or tens of kilovolts.

produce a field that is defocusing not only in the vicinity of the accel electrode, which is inevitable, but in the entire decelerating region as well. Under these conditions it is difficult to produce a well collimated ion beam.

In Fig. 5.15b the effect of a decel electrode placed closer to the accel electrode than the natural beam plasma boundary position is seen to lead to equipotentials which produce a focusing action. For situations where a beam with very narrow divergence is needed the three (or more) electrode system is desirable. For ion sources for electric propulsion, for example, the small loss in thrust associated with substantial divergence angles is acceptable and a two electrode system may be usable. For ion sources for neutral beam lines for mirror machines or Tokamaks highly collimated beam are required and a three electrode system is advantageous.

Finally a decel electrode will intercept a substantial fraction of charge transfer ions which drift back to the source and cause power loss and erosion. All things considered, the relatively slight complexity of the third electrode usually is warranted.

As pointed out, the use of a large accel voltage to beam voltage ratio makes possible a coarser geometry (or higher current density for a given geometry) than would be possible if this ratio is close to 1. However, there can be deleterious consequences of large accel–beam voltage ratios as well. These arise from charge transfers between neutral gas atoms and the beam ions, which will be discussed in detail in Sec. 5.6. The consequence of such charge transfers is to change fast ions into fast neutral atoms and slow gas atoms into slow ions. As a consequence of these reactions within the beam plasma this plasma contains slow ions. Those that drift back to the plasma boundary facing the source must be accelerated into the accel electrode. If the decel voltage (see Fig. 1.1 for definition of decel voltage) is large, then this current results in a loss of power and heating of the electrode, which is worse than if it were minimal. Ions that strike the decel electrode at high energy are also more likely to erode this electrode by sputtering (ejection of electrode material under ion bombardment) and will, until very high energies are reached, produce more secondary electron current. Those secondary electrons produced on the upstream portion of the electrode are accelerated to the source, thereby further increasing the power loss and drain.

These arguments relate only to the slow charge transfer ions produced in the beam downstream of the decel region. Those produced within the region between the source and beam plasmas can be even more harmful. The electrodes can be well designed for the ions originating in the source plasma but those that originate in the acceleration region will not follow the desired paths. A certain fraction of them will be accelerated through the electrodes at larger angles than the system is designed for, to produce parasitic beams studied in detail by Anderson (1978). Others will be accelerated into the accel electrode and the greater the ratio of accel–beam voltage the greater will be the fraction of charge transfer ions which strike this electrode and the greater will be the energy with which these ions will strike this electrode. Again these conditions usually make it advantageous to operate with a minimal decel voltage.

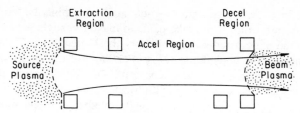

Figure 5.16. Four electrode extraction systems.

Kim et al. (1978) advance arguments favoring the four electrode system seen in Fig. 5.16, the extraction–accel–decel system. The advantage of such a system is particularly great when ion energies of the order of 100 keV or greater are desired. This relates to experimental evidence that the breakdown voltage across a gap is approximately proportional to the square root of the interelectrode spacing. With a current density limited by space charge to some constant times $V^{3/2}/d^2$ this would lead to an available current for a three electrode system proportional to $V^{-5/2}$, for voltages so high that the accel gap was limited by sparking. This limitation disappears for the extraction–accel–decel system where the extraction voltage and acceleration voltage are independent.

There are several other advantages as well. The ratio of the aperture diameter to the length of the extraction gap could be made larger than the value of 1.0 found by Coupland et al. (1973) to be a maximum for low beam divergence for a three electrode system. Kim et al. (1978) also found that this electrode system was capable of yielding a lower beam divergence and that the beam divergence was less sensitive than three electrode systems to variations in the source plasma density.

In addition, it proved possible, as shown by Menon et al. (1985), to design the electrode system seen in Fig. 5.20 such that slow ions from the beam plasma are forced into intercepting the last electrode and do not strike the accel electrode (#3) where they can create secondary electrons which can go back to the source where they produce significant problems associated with absorbing the large amounts of power.

5.6 Extraction Electrode Design

Cooper et al. (1972a) demonstrated that in a multiapertured electrode system each aperture acts independently, that is, the perveance per aperture and the angular divergence are the same as calculated and obtained for a single aperture, even when the apertures are packed close together. Thus the design of an electrode system can focus on a single aperture.

If the purpose of an electrode system is to produce a moderately high perveance (poissance per aperture of the order of 0.10 or higher) and virtually no direct beam interception, but without much concern for the angular divergence, then it is simple to design an electrode geometry. That shown in Fig. 5.15b, for example, drawn without reference to any computation, would almost certainly be satisfactory. Note that in drawing this figure, I did take pains to make the angle between the plasma electrode and the anticipated outer trajectory the Pierce angle 67.5°. This is desirable to minimize the nonlaminar (crossover) flow of the outermost ions, but it is certainly not essential, and many successful electrode systems have had holes in the electrodes which were purely cylindrical.

If, in addition to high perveance and negligible beam interception, it is necessary to produce a beam of very small angular spread, say of the order of ±1° or smaller, then one cannot be cavalier about designing the extraction electrode system. It is necessary to go through an extensive program of varying electrode geometry until an optimum is found. This may be done through an experimental program in which the operational parameters, including beam divergence angle, are measured as a function of various geometric parameters, or alternatively by a computational program.

An experimental program of this sort was carried out by Coupland et al. (1973) working with the electrode system shown in Fig. 5.17. The most critical parameter was found to be the ratio of the aperture radius r_1 to the electrode spacing d_1. They obtained maximum current density for a given voltage, and minimum angular spread, for $r_1/d_1 > 0.5$, that is, when the hole diameter was less than the spacing between the first two electrodes. The beam parameters were insensitive to r_2/r_1 but it was necessary to make this parameter > 0.75 to avoid impact of the beam on the accel electrode. The radius of the exit aperture r_3 appeared to have no effect. Coupland et al. found it of some importance to make the deceleration gap d_2 as small as possible, and this probably relates to the focusing action discussed in Sec. 5.4 and illustrated in Fig. 5.15b. Coupland et al. also found that the thickness D_1 of the plasma electrode had an effect, with an optimum at D_1 approximately $0.5r_1$.

They also investigated the effect of the negative voltage (the decel voltage) V^- on performance and found that this voltage had to be greater than $0.03V^+$ to pre-

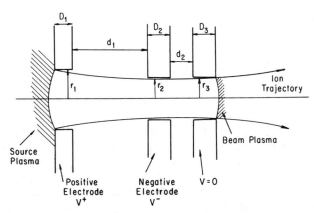

Figure 5.17. Extraction electrode geometry studied experimentally by Coupland et al. (1973).

vent electron backstreaming. For their geometry increasing V^- beyond this value had very little effect on the current into a small angular acceptance collector.

Another approach to electrode design is through the use of electrolytic tanks, as presented by Hyman et al. (1964). However, this method is so clearly superseded by computational techniques that it does not warrant discussion here. In one computational approach an electrode geometry and set of potentials is chosen. Then shapes of the source and beam plasma boundaries can be guessed at. As a start one might find the solution $V(x, y, z)$ of Laplace's equation for these boundaries. The next step would be to compute the trajectories of ions which start from rest at the assumed source plasma boundary. Then putting in a current density to this boundary yields a value of the space charge density ρ at all points. Based on this $\rho(x, y, z)$ a new $V(x, y, z)$ must be found as a solution of Poisson's equation. It is then necessary to rerun the trajectories and iterate until a self-consistent set of trajectories and potentials is obtained. The current density to the plasma boundary must then be readjusted in such a way that the final solution corresponds to a potential gradient very close to zero over the plasma boundaries. The measure of whether the assumed plasma boundaries are realistic is the uniformity of the current density over the plasma boundary, and the measure of whether the electrode shapes and source plasma boundary are satisfactory is the character of the beam as it passes the last electrode.

Cooper et al. (1972a) carried out such programs, iterating until they achieved a set of trajectories parallel to within $+1°$ for the circular aperture geometry shown in Fig. 5.18 and for the slot geometry shown in Fig. 5.19. The measured performance of their extractors matched the calculated performance quite well. The beam had a gaussian shape with a $1/e$ width of $\pm1.23°$ as a single aperture collector and $\pm1.31°$ when used in a 40% transparent array of holes. The calculated perveance was $6.5 \times 10^{-9}\ \text{A}/\text{V}^{3/2}$ for D^+ and $4.6 \times 10^{-9}\ \text{A}/\text{V}^{3/2}$ for D_2^+ so that the measured perveance of 5.1×10^{-9}, or 5.9×10^{-9} per hole when a 19 aperture extractor was used, fell in between. The exact composition of the beam was not measured in these experiments, but the results must be considered to be an excellent verification of the validity of the computer program.

The perveance quoted, $6.5 \times 10^{-9}\ \text{A}/\text{V}^{3/2}$ for D^+, corresponds to a poissance

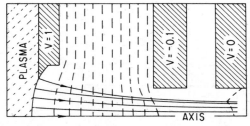

Figure 5.18. Electrode geometry of Cooper, Berkner, and Pyle for a circular aperture showing calculated ion trajectories and equipotentials.

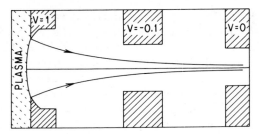

Figure 5.19. Electrode geometry of Cooper, Berkner, and Pyle for a long slot showing electrode potentials and trajectories of the outermost ions.

value of 0.17 and is 0.49 times as large as the perveance P_0 calculated on the basis of a Child equation current between the plasma surface and an electrode spaced from it by the leading edge of the accel electrode.

A more sophisticated program has been developed by Wheatley and co-workers at Oak Ridge National Laboratory, and it is described in an extensive series of papers which are enumerated in the ion source description of Menon et al. (1985). Wheatley's computational technique enables him to start the computation not at an assumed plasma boundary with ions at rest but actually well back in the plasma so that the velocity given the ions by the weak fields within the plasma toward the electrodes are taken into account. The application of the developed program to the four electrode system shown in Fig. 5.20 yielded a beam whose angular spread was only ±0.26°. The results of this calculation (shown in Fig. 5.20) were found consistent with observations made on the beam.

The unusually thick ground electrode served an important function. Its depth, plus the shape of electrode number 3, caused most slow ions from the beam plasma to be intercepted before being caught in the field between electrodes number 3 and 4. This in turn minimized the production of secondary electrons at electrode number 3.

Poissance values of less than 0.2 per aperture are the most that can be obtained for beams of very small angular spread. For ion engines where spreads as large as ±30° are tolerable much larger values can be obtained. For example, Nakanishi

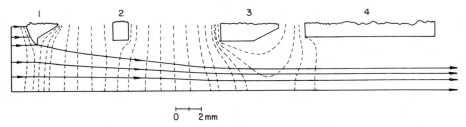

Figure 5.20. Electrode geometry of Menon et al. (1985) with computed trajectories and equipotentials.

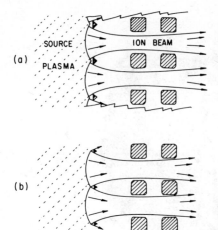

Figure 5.21. (*a*) Probable cause of drain reduction in narrow plasma electrodes (*b*) the possibility of the virtual elimination of the plasma electrode drain.

et al. (1968) quote perveances per hole for mercury of 2.5×10^{-9} A/V$^{3/2}$ which translates for Hg$^+$ ions to a poissance of 0.65. The presence of a small percentage of Hg^{2+} would lower this value somewhat, but it is certainly much larger than 0.2.

For electrode designs in which very small angular spreads are unimportant, as in ion engines, the plasma electrode can be made thin and the apertures can be pushed very close together to achieve high transparency electrodes. In this case it has been observed that the ion current to a cathode potential plasma electrode ′ decreased more than in proportion to the electrode area, as the space between apertures was decreased.* This is attributed to the effect illustrated in Fig. 5.21*a* in which the fields pull ions from a portion of the space behind the electrode. Near the edge of an extraction aperture the accelerating field can be expected to enlarge the sheath thickness. It ought to be possible to extend this idea to the limit shown in Fig. 5.21*b* with virtually no charged particles striking the plasma electrode.

For this idea to have any merit it is necessary that the sheath thickness have a size which is not an order of magnitude smaller than the plasma electrode thickness. For example, consider a plasma with a density of 5×10^{11}cm^{-3} and an electron temperature of 6 eV, a source capable of delivering, according to Eq. (3.69), 93 mA/cm^2 of H$^+$ ions. Suppose this plasma faces an electrode which is 60 volts negative. From Fig. 3.12 we see that the sheath thickness will be about 13 Debye lengths or 0.33 mm. It may be feasible, for plasmas of this density, to carry out the reduction in interception as indicated in Fig. 5.21. This reduction would not be for the purpose of achieving the small resultant increase in extracted current but rather to decrease the power delivered to the plasma electrode and, as a result, increase its mechanical stability without the necessity of cooling.

*This is an observation made by co-workers and me at Electro-Optical Systems, Inc. around 1964. I can find no literature reference.

Figure 5.22. Ion engine of Sohl *et al.* (1966) showing the gradation in extraction hole size to compensate for the variation in plasma density across the source diameter.

One of the problems of ion extraction from an extended area relates to the variation in plasma density over the extraction area. This problem was attacked by Sohl et al. (1966) by a gradation in the extraction aperture diameter as shown in Fig. 5.22. The solution of King et al. (1967) was to make the accel electrode convex toward the screen electrode, so that the interelectrode spacing increased roughly as the square of the radius. Both solutions proved adequate for ion engine applications where angular divergence is of little consequence until it becomes large enough to significantly decrease thrust. For applications where angular divergence must be of the order of $\pm 1°$ or less it is necessary to produce plasmas with a highly uniform profile. The various means of achieving this uniformity are examined in Chapter 8.

5.7 Effect of Neutral Vapor

It is always important to make the gas utilization efficiency of an ion source as high as possible. Some of the reasons are very clear. If the ion source is meant for space propulsion, then neutral gas represents mass which must be carried into space from which negligible thrust is derived, and this degrades the performance. In the laboratory neutral gas increases the requirements for pumping, and in some applications it leads to the requirements for gigantic cryopanels to obtain the pumping speed required to adequately handle the gas.

Of greater importance, usually, are the consequences of charge transfer between the ions and neutral atoms or molecules, especially when it occurs in the electrode region. When such an interaction occurs between a fast ion and a thermal velocity neutral particle the result is a fast neutral and a slow ion. With an electrode system designed to form a well defined beam out of those particles formed inside the plasma which drift to the boundary, the ions generated in the acceleration region are likely to bombard the electrode or to leave the source at wide angles. The fast neutrals, of course, are not controlled by the fields and diverge to form parasitic components as described by Anderson (1978), which, for high energy beams, may create problems by delivering power and sputtering agents to undesirable regions of the ion beam duct.

The effects of electrode interception of fast ions are apt to be the most harmful. For example, in D^+ sources such as required for the generation of energetic neutral beams one aims at current densities of 0.25 A/cm^2 at, say, 150 kV. It would not require a very large fraction of this 37.5 kW/cm^2 to melt or vaporize the accel electrode and even 10^{-4} of that power would cause the electrode to run very hot. Furthermore, according to data of Barnett et al. (1977), each ion which strikes the electrode at this energy will release, on the average, two secondary electrons and the handling of power delivered to the plasma electrode and the ion source generally by backstreaming electrons is one of the serious problems of high energy D^+ ion sources.

Another serious consequence of electrode bombardment is erosion of the electrode by sputtering. For light ions such as H^+ or D^+ the sputtering yield (atoms per ion) is generally small and by a proper choice of electrode material can be made very small. For example, Barnett et al. (1977) show a maximum sputtering yield for D^+ on tungsten of about 10^{-3} at 4 keV. On the other hand, where heavy ions, such as Cs^+, Hg^+, or Xe^+ are taken to about 5 kV, as in ion propulsion, sputtering yields may be large, of the order of 3 even for the most favorable electrode material. For example, if only 10^{-6} of a 1 mA/cm^2 beam is caused to strike an electrode for which the sputtering yield is 1, it is easy to show that in the course of one year (a short time compared to times envisioned for some space missions) 30 g of molybdenum electrode per square centimeter would be sputtered away.

As an example consider a hydrogen ion source generating 0.2 A/cm^2 of which 60% are H^+ ions and 40% H_2^+ ions and suppose that the flow of neutral gas approximately matches the ion flow. (These are moderately realistic numbers for some hydrogen ion sources.) How much charge exchange occurs in a 1 cm distance in which the ions get accelerated to 10 kV?

If we suppose the neutral gas to have a temperature of about 0.05 eV (580K), we get $v_H \approx 3 \times 10^5$ cm/s and $v_{H_2} \approx 2 \times 10^5$ cm/s. For the flow figures given these yield densities $n_H \approx n_{H_2} \approx 2.4 \times 10^{12}$ cm^{-3}. As an approximation use the following average values of the charge transfer cross section obtained from Fig. 5.23:

$$H^+ + H, \qquad \sigma = 2 \times 10^{-15} \text{ cm}^2$$

$$H^+ + H_2, \qquad \sigma = 4 \times 10^{-16} \text{ cm}^2$$

$$H_2^+ + H, \qquad \sigma = 10^{-15} \text{ cm}^2$$

$$H_2^+ + H_2, \qquad \sigma = 6 \times 10^{-16} \text{ cm}^2.$$

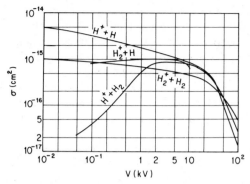

Figure 5.23. Charge transfer cross sections for H^+ and H_2^+ ions in atomic and molecular hydrogen.

The fractions of ions which charge exchange, σnl, are then

$$H^+ + H_2, \qquad 4.8 \times 10^{-3}$$

$$H^+ \text{ in } H_2, \qquad 1.0 \times 10^{-3}$$

$$H_2^+ \text{ in } H, \qquad 2.4 \times 10^{-3}$$

$$H_2^+ \text{ in } H_2, \qquad 1.4 \times 10^{-3}.$$

For these numbers we see that about 0.5% of the ion beam is expected to undergo a charge transfer collision in the 1 cm distance. It may sound like a small amount, but it is the primary source of drain currents and electrode heating.

In other materials the charge transfer cross sections may be much greater. For cesium ions in cesium vapor, for example, the cross section varies from about 5 $\times\ 10^{-14}$ cm^2 at 50 eV down to about 10^{-14} cm^2 at 10 keV. Fortunately, for applications requiring ions with such large charge transfer cross sections the ionization cross sections are also high so that the neutral vapor out of the source can be made much less. In all cases, however, minimizing the neutral vapor from a source is an extremely important consideration.

In the preceding discussion the harmful effects of charge exchange in the accelerating gap were emphasized, but the production of slow ions downstream of the electrode is also harmful. This puts slow ions in the beam plasma and the slow ions which drift to the decel electrode can be accelerated to the accel electrode where they produce secondary electrons, which in turn are accelerated to the source.

5.8   Beam Steering

It is sometimes important to be able to vary the beam direction, as for thrust vectoring of ion engines. For small diameter, low poissance beams one can easily deflect the ions with a crossed electric field. For a broad high poissance beam

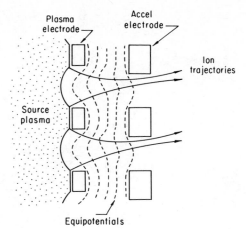

Figure 5.24. Effect of a lateral displacement of the accel electrode on beam direction. (The ion trajectories and equipotentials are *not* calculated curves.)

such as generated by a multiaperture electrode system this cannot be done. The beam plasma in that case will simply form sheaths that will prevent penetration of the fields into the beam. Beam deflection can be obtained by a lateral deflection of the accelerating electrode, as shown in Fig. 5.24 for a two electrode system. A small lateral deflection of the electrode in one direction causes a deflection of the beam in the opposite direction. In good electrode systems, as seen in Figs. 5.18–5.20, the edge of the beam is far enough from the electrode so that small lateral displacement of the electrode is feasible before encountering beam interception. Sohl and Fosnight (1969) and Fosnight et al. (1970) built an ion source in which the accel electrode could be laterally displaced by electrically controlled thermal expansion of support rods. They were able to obtain beam deflections of ±9° for two orthogonal directions for very small increases in the current to the accel electrode.

5.9 Insulator Coated Electrodes

As described in Sec. 5.5 and illustrated in Fig. 5.15a, an effective extraction and acceleration system might consist of only two electrodes, although low angular divergence has been obtained only with three or more electrodes. An electrode system which uses only a single electrode can be made by applying an insulating layer to the plasma side of the screen electrode and using the screen electrode as the accelerating electrode. This was first tried by Margosian (1967), who sprayed alumina on the back of an electrode, but more favorable results were obtained by Banks (1968) and Nakanishi et al. (1968) by bonding glass to the upstream side of a metal grid as shown in Fig. 5.25.

The portion of the glass in contact with the plasma must come to the floating potential relative to the plasma, that is, the potential at which the ion and electron currents are equal. According to Fig. 3.11 this would be about six times the plasma electron temperature for a mercury plasma containing only maxwellian electrons.

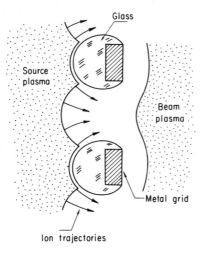

Ion trajectories

Figure 5.25. Glass coated extraction electrode.

For the mercury plasmas actually used a small percentage of primary electrons would make the glass surface somewhat more negative relative to the plasma. The metal grid will be made at a still more negative potential providing a very compact, high perveance extraction system with only a single electrode. Alignment problems disappear.

These grids, in fact, exhibited higher perveance per aperture than otherwise obtainable. Performance was found by Bechtel et al. (1971) to be improved by sputtered material which coated the portion of the glass visible from the downstream direction. This suggests that the metal grid, seen in Fig. 5.25 to be coated by glass on all except the downstream surface, ought to be coated only on the side facing on the plasma. This type of grid presumably produces a fairly broad angular divergence, but this property of the beams has not been reported as far as I could determine.

The shortcoming of the coated single grid system that finally led to discontinuation of the system was its relatively short lifetime. The pursuit was carried out for possible applications to propulsion where lifetimes of the order of 10,000 hours are required. These electrodes generally failed after a few hundred hours at most.

However, the use of a coated screen grid as part of a three electrode system was subsequently studied. With a layer of Al_2O_3 on the plasma side of a screen grid with cylindrical holes, Whealton, Grisham, et al. (1978) were able to substantially improve angular divergence. For example, with an accel voltage of 14.4 kV the use of a 300 volt potential on the insulated plasma grid reduced the angular divergence from $1.16°$ to $0.55°$ at the same current. Consistent with these results Whealton, Tsai, et al. (1978) showed that the transmission from a 22 cm diameter source through a 20×25 cm aperture 4.1 m downstream was increased 30% by the use of an insulated coating on the first grid. As favorable as these results appear, they seem to have been abandoned in favor of four electrode systems with shaped holes in the plasma electrode.

5.10   Fine Mesh Ion Extraction

It is feasible to extract an ion beam by having a single fine mesh at a negative potential face the source plasma as described by Forrester et al. (1975), Crow et al. (1978), and Harper et al. (1980). If the hole size in the mesh is sufficiently small relative to the sheath thickness S_0 the acceleration of the ions is essentially between parallel planes.

In Fig. 5.26 the source potential is shown to be V_0 relative to the beam plasma potential. The accelerating grid is at $-V_1$ to provide the accel–decel potential variation required to prevent a backflow of electrons. Actually, a bias of the grid relative to the channel through which the beam must travel is not necessary. With the grid at the same potential as the beam surroundings the beam potential will automatically rise to the level necessary to trap the electrons required for neutrality.

At the beam edge Fig. 5.26 shows electrodes which approximate the Pierce geometry. In practice this has not proven necessary for the accel electrode, the electrode holding the grid, and the grid can simply be made to cover a larger area than the circular or rectangular aperture in the source plasma boundary.

I have also shown the distance S_0 in Fig. 5.26 to be such that the source plasma–sheath boundary facing the extraction grid is coplanar with the sheath boundary behind the aperture defining electrode. Departures from the current voltage relationship required to produce this condition which decrease S_0 will cause a divergence of ions at the beam edge and, conversely, increasing S_0 will lead to crossover ions from the beam edge. This may involve only a small fraction of the total current for sources in which S_0 is of the order of 0.3 cm and the aperture is, say, a square 10 cm on a side, but this can be a nuisance nevertheless. One serious effect encountered when S_0 is too large is a focusing of ions to a line on the grid which may sputter through rapidly. To avoid such difficulty it is necessary that for

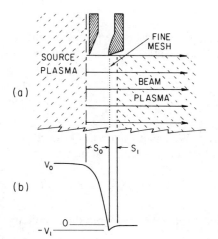

Figure 5.26. Fine mesh extraction system (*a*) and potential variation (*b*) from the source to the beam plasma.

each current density generated the appropriate voltage be used. Setting the correct voltage is easy if the beam can be observed visually. Otherwise the correct voltage for each current density can be obtained from the material covered in Sec. 3.8. (See, for example, Problem 3.5.) Independence of current density and voltage can be avoided by the use of two fine grids and I shall return to that possibility later. Before leaving single grids it is important to examine the angular spreads and the various limitations for this system.

A. Angular Divergence

For this acceleration system there are two conceivable causes of angular divergence, aside from that due to lateral velocities which the ions might have as they arrive at the extraction sheath. An angular spread can result from a spatial ripple in the plasma boundary reflecting the periodicity in the grid, and divergence can be introduced by the transverse components of the field in the vicinity of the grid wires.

Consider the field near a grid of characteristic spacing λ in two perpendicular directions. In a vacuum situation the slowest varying term in a Fourier expansion of the potential would vary as $\exp(-2\sqrt{2}\pi x/\lambda)$ where x is distance away from the grid. Space charge effects should produce some shielding, which would cause the ripple in potential to fall off even faster. If x/λ is as small as 1.5, this term is 1.6×10^{-6}. We can probably neglect departure from flatness of the plasma surface whenever the ratio of the distance S_0 to the grid wire–to–wire spacing is 1.5 or greater.

An approximation to the divergence caused by fields near the grids can be obtained by considering each aperture to act independently and use Eq. (5.14) for the focal length of an aperture. Then take the field on the downstream side as negligible, obtaining

$$f = -4V_g/V_g', \tag{5.22}$$

where V_g is the voltage between the plasma and grid and V_g' is the potential gradient at the plasma side of the grid. From Fig. 5.27 we see that this leads to an angular divergence of the outermost ions given by

$$\theta = b/2|f|, \tag{5.23}$$

where b is the effective diameter of an aperture. We can estimate V_g'/V_g occurring in Eq. (5.22) for f from the Child equation obtaining $f = 3S_0$ leading to

$$\theta = b/6S_0. \tag{5.24}$$

The material of Sec. 3.8 would yield a smaller value of V_g'/V_g but the simplicity

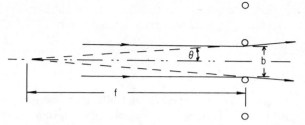

Figure 5.27. Illustration of divergence caused by passage of a parallel ion beam through a grid.

of this analysis suggests this approximate route. In terms of the current density and voltage across the gap we obtain

$$\theta = \frac{b}{6} \sqrt{\frac{J}{\chi}} \, V_g^{-3/4},$$
(5.25)

where χ is the Child equation constant $(4\epsilon_0/9 \, \sqrt{2e/M}$. In Fig. 5.28 J is plotted as a function of V_g with θ as a parameter for the case of H^+ ion extraction from a plasma with a woven grid made of wire 2.5×10^{-3} cm in diameter on a 55 wires per centimeter mesh, a readily available rugged material. Woven tungsten grids finer than this are available and still finer photoetched grids can be obtained. Nickel mesh very much finer can be obtained but because of power and sputtering limitations tungsten is the material of choice for this application.

B.   Power Limitation

Unlike electrodes in which carefully aligned apertures cause the ion beam to pass through the accel electrode without interception, the fine mesh accelerator is as-

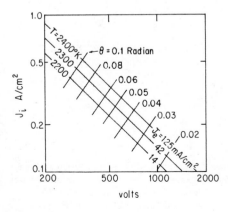

Figure 5.28. Current–voltage curves for various temperature and angular beam spreading for H^+ ion beams accelerated by a 55 cm^{-1} mesh made of 2.5×10^{-3} cm diameter tungsten wires.

saulted directly. If the grid is made of circular wires bombarded on one side, but radiating in all directions, the grid will reach a temperature T, in Kelvins given by

$$JV = \pi e\sigma T^4, \tag{5.26}$$

where e is the emissivity and σ is the Stefan–Boltzmann constant. The clear preference for the grid material is tungsten. Not only can tungsten be woven and photoetched into very fine grids but the temperature to which it can be raised is the highest of any material. In addition, it has a low sputtering yield compared to most metals. We shall therefore consider only tungsten as the grid material.

The emissivity of tungsten, shown in Fig. 5.29, is seen to be approximately linear and given by

$$e = 1.25 \times 10^{-4}T \tag{5.27}$$

for 500 K $< T <$ 2500 K. The temperature limitation is not the melting point of tungsten but rather the temperature at which the grid became so electron emitting that the power supplies cannot accept the currents or the ion source the heating power. This limit would be about 2400 K where the emission current density J_e for tungsten, as shown in Fig. 5.28, is about 125 mA/cm^2. We can see from Fig. 5.28 that very substantial low energy ion beams can be obtained. For H$^+$ ion beams at, for example, 600 volts, even with a grid temperature limited to 2300 K it is possible to obtain a current density from the plasma of 0.24 A/cm^2 with an angular spread of ±0.045 radians $= \pm2.6°$. Grids such as that for which Fig. 5.28 are plotted are typically 70% transparent so that the extracted current would be 0.17 A/cm^2. Such performance was routinely observed in my laboratory at UCLA.

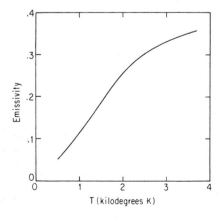

Figure 5.29. Total emissivity of tungsten as a function of temperature, from data of Roeser and Wensel (in *Handbook of Chemistry and Physics*, 65th ed., CRC Press, Boca Raton, FL, 1984, p. E-381).

C.   Sputtering Limitations

Ions striking a solid surface eject some of the material of the solid in a phenomenon known as sputtering. Once again tungsten appears to have a particularly low sputtering yield and is the recommended material for this application. Sputtering may not be a serious problem when the bombarding ions are H^+ because the sputtering yield is small for light ions. For example, the sputtering yield for 1 keV H^+ ions striking a W surface at normal incidence is given by Finfgeld (in C. F. Barnett et al., 1977) as 2.3×10^{-4} atoms/ion, and falling off as energy is decreased. This would lead to operation for 122 hours before a 0.1 A/cm^2, 1 kV H^+ beam would erode tungsten to a depth of 10^{-3} cm.

For heavier ions the sputtering yields are much less favorable. Consider for example an A^+ ion beam incident on the same 55 cm^{-1} tungsten mesh. From data of Henschke and Derby (1963) and Carlston et al. (1965) I estimate that the sputtering yield of A^+ ions on tungsten can be approximated by

$$\gamma = (1.65)(10^{-3})(V - 75),$$

for V in the range of 100–1000 volts. If we take 10 μm = 10^{-3} cm as a standard measure of depth of grid erosion, then we can readily show that the running time before this much erosion occurs is given by

$$\tau = \frac{1.701}{J_i(V - 75)},$$

where J_i is in amps per square centimeter and V is in volts. J_i as a function of the source to grid voltage is shown in Fig. 5.30 for various times required to erode to a depth of 10 μm. The angular divergence obtained from Eq. (5.25) is also shown in this figure.

D.   Contoured Fine Mesh

Although the fine mesh does not have the rigidity to be given a permanent spherical shape it can be given an approximation to a spherical shape by deforming a circular grid with one or more rings pressed into the mesh as shown in Fig. 5.31. The configuration shown was successfully used in my laboratory to produce a beam narrowing relative to the source aperture.

E.   Double Grid System

Crow et al. (1978) used two parallel fine meshes to form a system in which the ion current density and accelerating voltage are independently variable. Such a system is illustrated in Fig. 5.32. An interesting phenomenon was observed with the two grid accelerator. Ion current can be extracted which is several times the

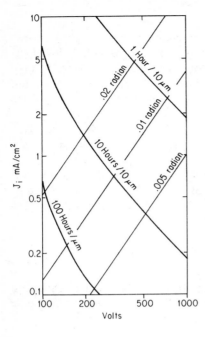

Figure 5.30. Current–voltage curves for various times to erode a 10 μm thickness, and angular beam spreading for A$^+$ ion beams accelerated by a 55 cm^{-1} mesh made of 2.5×10^{-3} cm diameter tungsten wires.

calculated space charge limited flow for a given accelerator voltage and grid spacing. The formation of a potential maximum between the screen and accel grid which would be large enough to turn ions back is prevented by the trapping of electrons in the maximum, forming a plasma slab. Calculated potentials for a fixed current and various accel voltages are shown in Fig. 5.33.

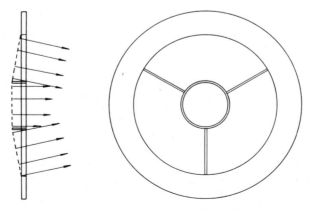

Figure 5.31. Technique for contouring fine mesh extractors.

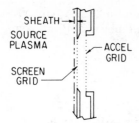

Figure 5.32. Double grid accelerating system.

F.   Application Hints

Within the range in which fine mesh extraction is applicable it is capable of producing high density, high uniformity beams of ions with very small divergence angles. To achieve these features it has been found necessary to utilize springs which keep the grids under tension. Even a very small waviness is reflected in nonuniformities in the beam. With the two grid system it is also important that the two grids have substantially different spacings to avoid moiré patterns, which can seriously degrade the beam uniformity.

It is extremely important to avoid the phenomenon of spotting, discussed in Sec. 6.1, when using a fine mesh extractor because of the fragility of the grid. Sparking to the grid often results in a puncture. To minimize spotting the vacuum system should be extremely free of contaminated surfaces and a plasma bridge neutralizer should be used. As mentioned in Sec. 6.1, the use of a cesiated plasma bridge neutralizer is suggested.

G.   Not So Fine Single Grid Extraction

Single grid extraction can be useful in ranges where *fine* mesh might be an inappropriate term. For example, Delaunay et al. (1983) use a single grid extraction

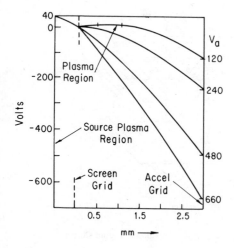

Figure 5.33. Calculated potential distributions for a two grid system for A^+ ions with a current density of 0.9 mA/cm^2.

system with 0.5 mm diameter apertures for the extraction of a 20 mA/cm², 0.5 eV H⁺ ion beam. The reader is given the opportunity, in Problem 5.10, of computing the angular spread of the extracted ions.

5.11 Magnetic Double Sheath Acceleration

It is possible to create a virtual grid to accelerate ions from a surface at which they are produced, or to which they are delivered, so that ions can be extracted from a broad area across a narrow sheath without either a multiaperture acceleration electrode system or a fine mesh. Such a virtual planar electrode can be established by a Hall current flow of electrons in a magnetic field parallel to the surface from which ions are to be extracted.

The sheath configuration is illustrated in Fig. 5.34 and the nature of the potential variation is shown in Fig. 5.35. Ions start from the left in both figures, and are accelerated to the right. The deflection of the ions in the magnetic field will be small and has been neglected. At the right side of the sheath, the surface labeled cathode boundary, electrons are released and accelerated to the left. For the correct magnitude magnetic field these are constrained to move in cycloidlike orbits of a limited excursion, progressing in the $\mathbf{E} \times \mathbf{B}$ direction as shown in Fig. 5.33. The region to the left of the space in which electrons gyrate contains only ions and is

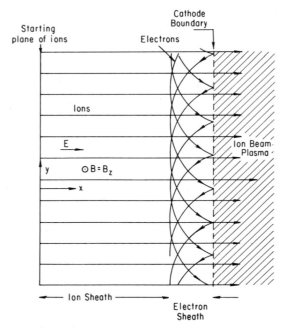

Figure 5.34. Geometry and particle trajectories for the magnetic double sheath.

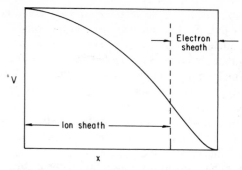

Figure 5.35. Potential variation through the magnetic double sheath.

termed the *ion sheath*. It can be shown that within the portion of the sheath containing electrons as well as ions the electron space charge predominates and this region is termed the *electron sheath*.

The double sheath has been shown by Sudan and Lovelace (1973) and Forrester (1975) to be subject to an analytical treatment by matching the solutions of Poisson's equation in the ion and electron sheaths. In particular one may choose zero potential gradient at the two sheath boundaries and in terms of the desired ion current density and total accelerating potential derive the required electron current density to and from the cathode boundary, the net Hall current of electrons, the ion and electron sheath thickness, and other details of the sheath which correspond to any magnetic field intensity.

The first source to make use of acceleration across such a double sheath was that of Meyer (1967, 1970), who computed his electron trajectories by numerical, rather than analytic techniques. His source had the annular symmetry shown in Fig. 5.36. Ions in his source were created at the planar surface of the porous tungsten, as described in Chapter 9. His magnet created a radial field which was

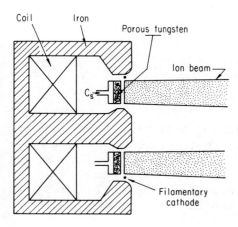

Figure 5.36. Space charge sheath electric thruster of Meyer.

parallel to the emitting surface in the acceleration gap. The acceleration plane (the cathode boundary of Fig. 5.34) is created by electrons which readily flow inward from the emitting filament along field lines. The cycloidal-type motion of the electrons is in the azimuthal direction. This may at first sound ideal since the electrons circulate instead of making a single pass across the gap but there is an overwhelming drawback. The type of current shear in this space charge sheath is subject to the diocotron instability. As long as each electron gets across the gap in a single pass made of a small number of hops, the instability does not have time to build up to a significant level. Where the electrons circulate this instability is almost certain to develop. This is probably the cause of the very large drain currents observed by Meyer, which made his source impractical for its intended use as a space thruster. Nevertheless, a low divergence, high intensity ion beam indicated that he did succeed in forming an effective double sheath acceleration system.

At present this principle is widely used in very high voltage, high current pulsed sources such as described by VanDevender and Cook (1986). Such sources are known as magnetically insulated diodes. The description and analysis of these is outside the scope of this book.

Problems

Section 5.2

5.1 A circular aperture 6 mm in diameter is cut in an electrode bounding a plasma. Another electrode spaced 6 mm from this electrode also has a 6 mm diameter hole. Minimum angular divergence is obtained when the current flow is only 0.4 time as great as that obtainable on the basis of the Child equation, that is, when the quantity P/P_0 defined by Eq. (5.5) is 0.4. What is the current of Ne^+ ions which can be obtained at 10^4 volts under conditions of minimum angular divergence? What is the poissance of that beam? (Atomic mass of Ne = 20.)

Section 5.3

5.2 A Pierce electrode system is used to accelerate a very low energy, singly ionized argon ion beam of current density 50 mA/cm^2 to 10^4 eV. If the beam is 1 mm wide, and long compared to its width, what angular divergence will the beam have as it emerges from the acceleration slit?

5.3 A circular aperture 0.7 cm in diameter in the boundary of a cesium plasma is surrounded by Pierce electrodes such as shown in Fig. 5.7 calculated to produce a parallel flow of ions between the electrodes. The distance between electrodes is 1.5 cm. If the potential difference between the electrodes is 5 kV, how much current uniformly delivered to the aperture will result in the desired parallel flow? What is the poissance of this electrode system?

5.4 Consider the emitter shown in Fig. 5.9 to be a cylindrical strip of length b $\gg a$, and suppose the electrodes are those shown in Fig. 5.9 chosen to produce radial flow toward the axis of the cylinder. Find an expression for the poissance of this electrode system and evaluate for an emitter which subtends an angle $\theta = 30°$.

5.5 Suppose the emitter shown in Fig. 5.10 is a circular piece of a sphere. Find the poissance and evaluate for $\theta = 30°$.

Section 5.5

5.6 A typical mercury ion source operates with a central plasma density of 5×10^{11} electrons/cm^3 and an electron temperature of 4 eV. It is desired to obtain a beam of 1 A at 5 kV from such a plasma. Assume a simple two electrode extraction system with a poissance per aperture of 0.25 and a plasma grid transparency of 0.7. What is the required diameter of the extraction area? How many apertures are required? What is the diameter of an individual aperture?

5.7 Regard the 1 A ion beam of Fig. 5.6 as having a uniform current density of ions flowing parallel to the axis. Find the stalling distance in the absence of neutralization.

Section 5.10

5.8 In a fine mesh extraction system the grid is 3 mm from the surface of the beam defining aperture facing the source plasma (see Fig. 5.26). The plasma is an A^+ ion plasma with an electron temperature of 5 eV. What current density is required to produce a plasma–sheath boundary which is planar, as shown in Fig. 5.20, if the defining aperture electrode is 50 volts and the grid 300 volts negative relative to the center of the plasma?

5.9 A single grid facing a plasma has an array of 0.5 mm diameter holes. If the current density of H^+ ions delivered to the plasma boundary is 20 mA/cm^2, what angular spread is to be expected in the beam if the plasma is 500 volts positive relative to the grid?

Chapter 6

Propagation of High Poissance Beams

6.1 Ion Beam Neutralization

In Sec. 2.11 it was shown that charged particle beams for which the perveance $I/V^{3/2}$ approaches the value $\chi = (4\varepsilon_0/9)\sqrt{2e/M}$ are subject to severe space charge spreading. If the beam has a long narrow cross section, then the allowed perveance value can be multiplied by b/a, the ratio of length to width. A dimensionless parameter, poissance

$$\Pi = (I/V^{3/2})/(\chi b/a) \qquad (2.116)$$

was defined and the criteria for a well directed ion beam to propagate without severe spreading can be expressed as

$$\Pi \ll 1,$$

in the absence of any neutralization of the space charge. As Π became large compared to unity, not merely beam spreading but beam stalling could be expected. Actual ion beams often greatly transcend this limit. For example, my co-workers and I have routinely generated 10 cm diameter D^+ ion beams of the order of 20 A at 600 volts. The value of Π for this beam is 3.5×10^4, a value which would lead to a stalling distance of about 3×10^{-2} cm, according to Eq. (2.133). Instead this beam actually propagated 3 m to the end of the chamber.

The reason is clear. The fields that produce the space charge spreading of positive ions will lead to the trapping of electrons, so that the region through which the ions pass is a plasma. More will be said about this plasma in Sec. 6.2. Let us consider here the processes by which electrons are introduced and kept in the beam.

To keep electrons in the beam it is necessary, first of all, to be sure that an accel–decel extraction system is used, as pointed out in Sec. 5.4. That is, the potential variation between the source plasma and beam plasma is required to go through a minimum, as shown in Fig. 5.13, deep enough to prevent the flow of electrons from the beam back to the source.

It is also necessary that the beam be prevented from having access to other high positive voltage which can extract electrons. Figure 6.1a, for example, shows an electrode arrangement in which electrons in the beam region can readily escape the beam by going around the electrodes to the source potential. The consequence of such a configuration is shown in Fig. 5.12 where space charge divergence of a low poissance beam is shown. In Fig. 6.1b the correct configuration is shown. The beam is surrounded by the potential of the decel electrode. If no decel electrode is used, as discussed in Sec. 5.4, then a configuration such as shown in Fig. 6.1c is satisfactory. In a situation such as shown in Fig. 6.1b or c, the beam plasma will assume a potential sufficiently positive relative to the channel in which it propagates to contain its electrons.

There are several sources of neutralizing electrons. At high energies the ions can produce electrons by ionizing the ambient gas. Since ionization by singly charged ions is roughly the same as for electrons of the same velocity, it requires high energy beams to produce much ionization, although some ionization can occur even when the ion velocity is below the threshold velocity for ionization by electrons. When the ion energy is too low to produce any ionization of the gas, secondary electron emission remains as a source of electrons. Since electrons become trapped it is not necessary that the source be very prolific but secondary emission can be a copious source and it is a source that exists down to low ion energies. For example, a 500 eV H^+ ion gives a yield of 0.2 electron per ion from a molybdenum surface. At 10 kV this yield would rise to 0.9. Thus there are always space charge neutralizing electrons emitted from the ion beam target and,

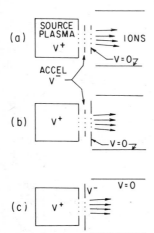

Figure 6.1. (a) Geometric arrangement which would permit electrons from the ion beam region to reach the source potential V^+ and (b), (c) geometries designed to prevent such electron escape.

in the case of the fine mesh accelerating system of Sec. 5.7, from the accel grid. If this grid is sufficiently negative relative to the beam region, these electrons can produce others by ionization of the neutral gas.

In summation, ion beams readily neutralize in a laboratory environment, although as described later it is sometimes advantageous to abet the introduction of electrons even in the laboratory. For ion sources contemplated for space propulsion the neutralization problem becomes somewhat different. Clearly a space ship cannot send out a positive ion beam without emitting at the same time an equal current of electrons. Failure to do so would cause the vehicle to become so negative that the ions could not escape the vehicle. The necessary zero current condition might be obtained by having an electron gun on board shooting out a current which matches the ion beam. In that case it has been questioned whether such electrons would mix well enough with the ion beam to provide adequate space charge neutralization. Otherwise stalling of the very high poissance ion beam might occur.

Initial attempts at laboratory solutions to the problem involved stringing electron emitters close to the ion beam so that the beam could draw an adequate current into itself as it went slightly positive. The measure of success in achieving both space charge and current neutralization would be the propagation of the beam to a collector, say, 2 m away to which the total current was zero, with equal positive and negative currents from the ion source and neutralizing electron emitters, respectively. The actual ion current to a collector could still be observed by power measurements and thrust measurements made at the collector.

The problem of placing emitters close to the beam was made difficult by the nature of the sources. To achieve the efficiencies and exhaust velocities required for propulsion required beams of heavy ions (Cs^+ and Hg^+ were the common ions for this purpose) at energies of a few kiloelectron volts. These ions are such potent sputtering agents that an electron emitter close to the beam would be rapidly eroded by even a small number of stray ions.

This problem was solved by the plasma bridge neutralizer developed by Ernstene et al. (1966, 1970). In its original application to the neutralization of Cs^+ ion beams the plasma bridge neutralizer had the form shown in Fig. 6.2. A small amount of cesium vapor introduced into a small heated cylinder, typically 0.64 cm in diameter and 3 cm long running at 600°C, turns the inside walls into excellent electron emitters.* A discharge forms between the beam as anode and the inside of the heated cylinder. The electrode labeled keeper is not an essential element for steady state operation but does help the discharge start. It is connected to a positive voltage through a sufficiently high resistor so that in operation it draws very little current.

The plasma inside the hollow cesiated emitter runs positive relative to the walls so that not only can electrons from the boundary flow out of the aperture but ions from this plasma flow out as well, providing a conducting plasma bridge between this cathode and the beam. The orifice was typically 0.013 cm in diameter and the

---

*A discussion of the properties of adsorbed cesium films is included in Chapter 9.

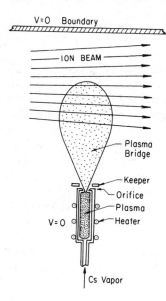

Figure 6.2. Plasma bridge neutralizer.

total rate of consumption of cesium by this cathode was very small. Only 2×10^{-3} to 10^{-2} as many cesium atoms and ions were emitted from this cathode as were electrons.

The cathode was maintained at ground potential, the same potential as the beam boundaries, and the beam ran at about 7 volts positive in order to draw electron currents which matched the ion currents, currents of the order of an ampere. Actually, cathodes of these dimensions can readily supply currents of 5 A for the amazing current density of 4×10^4 A/cm^2 from the orifice, and that does not appear to be a limit.

Although the hollow cathode is a particularly dramatic cathode when fed with cesium, insofar as it is a copious emitter, requires very low power, and has an unlimited lifetime, fearfulness of the effects of cesium have prevented its use with this vapor except in conjunction with cesium ion beams. However, even with other types of emitting surfaces and other gases or vapors it is a very useful neutralizer and has proven useful, for example, when the emitting surface was a typical BaO–CaO–SrO mixture and mercury vapor was fed through, or in another case when the emitting element was LaB_6 and the gas was hydrogen. These cathodes will be discussed further in Sec. 7.12.

As noted, the plasma bridge neutralizer was developed for space applications. In fact it has proven a very useful adjunct to the laboratory operation of ion beams. A short digression will help in the explanation. Ion beams are usually quite easily observed visually and if the beams are very faint the visibility can be improved by raising the background pressure. This is usually attributed to excitation of the ambient gas by the fast ions. This may well be the case for very high energy ions.

For low energy ions, say, 3 keV cesium ions, this is unlikely. The glow in the beam must be attributed to excitation by the more energetic of the trapped electrons. In the first instance where a plasma bridge neutralizer was used Ernstene et al. (1966) reported that when the discharge is struck to the plasma bridge neutralizer the blue glow characterizing the ion beam became very faint. It appears that the neutralizer supplied cooler electrons to the beam than the beam could obtain in other ways and that the energy of these was lower than required to excite the residual chamber gas.

I have observed another important effect of the use of a plasma bridge neutralizer in quieting phenomena in a beam. In running a source generating currents of 5–20 A of D^+ ions at 600 volts a poorly understood phenomenon of spotting is frequently observed—a phenomenon which appears associated with contaminated surfaces, such as accumulated pump oil. The phenomenon consists of a minute very bright spots, presumably electron emitting points, dancing around the chamber. This is accompanied by large amounts of electrical noise and, in spite of the fact that the highest impressed voltage is 600 volts, in spite of the fact that the low energy ion beam is inside an equipotential enclosure, the phenomena can be accompanied by violent sparking which frequently punctures the fine grids used to generate such beams. A hollow cathode used as a plasma bridge neutralizer is effective in quieting the observed phenomenon but not, unfortunately, in eliminating it. I believe that it would be greatly advantageous to use a cesiated plasma bridge neutralizer for the purpose, because cesium discharges make the coldest electrons of any discharge, and the minute amounts of cesium introduced into the chamber could easily, I believe, be handled with getters or refrigerant traps.

6.2 The Ion Beam as a Plasma

An ion beam always spreads more than can be accounted for on the basis of the angular divergence given the ions by the extraction electrodes. When this is observed it is almost always attributed to incomplete neutralization of the ion beam, a view which is basically incorrect. It is as if one stated that ions flow to the boundary of a plasma because the plasma is not completely neutralized. The fact that a plasma is positive in the center, permitting ions to drift out at the rate they are formed, is basic to a plasma. An intense ion beam exists in a plasma and the beam spreading is due to potential variations through the plasma and bounding sheath.

It is true that the potential variation in a plasma does correspond to a departure from perfect space charge neutrality, but the statement that the beam divergence is due to incomplete neutralization implies that if more electrons were made available to the beam the spreading would cease. In fact it is possible that supplying electrons can make the divergence worse. Furthermore, the statement suggests that the beam will spread like a beam of, say, only 0.01 times as much current. Instead the beam divergence, above that due to the angular spread with which the ions are injected, consists of a successive peeling off of the outer layers of the beam consisting of those ions which are in the sheath that forms at the beam boundary, as at the boundary of all plasmas, as shown by Crow (1977).

The picture Crow invokes is that of the ions moving through a plasma made of the beam ions themselves, slow ions formed either by ionization or by charge exchange between the beam ions and the neutral gas, and thermalized electrons. Under steady state conditions the potential distribution through the beam must assume a form such that ions and electrons must leave at the same rate they are introduced. At the charged particle densities characteristic of ion beams, the plasma can be considered collisionless and, following the Tonks–Langmuir treatment of Sec. 3.5, it will be assumed that ions formed in the beam are produced with zero velocity. Even when ions are formed by charge transfer this is a valid approximation since the charge transfer cross sections are very large compared to elastic collision cross sections. As in the treatment of a plasma without the fast ion component, the space between the bounding electrodes will be characterized by a region in which the deviation from neutrality is very small, that is, a region in which the plasma approximation $n_e \approx n_i$ is valid, and a sheath region where ion space charge predominates.

For simplicity I confine the analysis to a long slit beam of uniform fast ion density n_b. The cross section of the beam is of height b and width $2a$ with $b \gg 2a$, and the beam length l, the distance between the source and target, also satisfies $l \gg 2a$. I assume that slow ions are generated in the plasma at a rate g. Slow ions will have a velocity given by Eq. (3.30):

$$v(x_1, x) = \sqrt{2e[V(x_1) - V(x)]/M}, \tag{3.30}$$

where v is the velocity at the distance x from the median plane of those ions formed at x_1. Then

$$n_i(x) = n_b + \int_0^x \frac{g}{v(x_1, s)} \, dx_1, \tag{6.1}$$

which is Eq. (3.29) with the addition of the beam ion density n_b. In the plasma region we assume $n_i = n_e$ and

$$n_e = n_0 \exp(eV/kT).$$

The potential V is taken as equal to zero in the median plane, where the value of $n_e = n_0$, and T is the electron temperature. The plasma equation is then

$$n_0 \exp(eV/kT) = n_b + \sqrt{M/2e} \int_0^x g[V(x_1) - V(x)]^{-1/2} \, dx_1. \tag{6.2}$$

As before let

$$\eta = -eV/kT \tag{3.34}$$

and

$$\xi = x/L, \tag{3.34}$$

where L, given by Eq. (3.39), is readily seen from Eq. (3.37) with $\gamma = 0$ to be given by

$$L = (n_0/g)\sqrt{2kT/M}. \tag{6.3}$$

The plasma equation (6.2) then becomes

$$\epsilon^{-\eta} - \beta - \int_0^\xi (\eta - \eta_1)^{-1/2} \, d\xi_1 = 0, \tag{6.4}$$

where

$$\beta = n_b/n_0. \tag{6.5}$$

Except for the addition of the constant β, Eq. (6.4) is identical to Eq. (3.41) for the $\gamma = 0$ case. A treatment exactly parallel to that of Harrison and Thompson (1959), which led to solutions of Eq. (3.41), can be applied here, leading to

$$\xi = [F(\sqrt{\eta}) - \beta\sqrt{\eta}]2/\pi, \tag{6.6}$$

where

$$F(x) = \exp(-s^2) \int_0^x \exp(t^2) \, dt \tag{2.129}$$

is the Dawson function previously defined. For $\beta = 0$, Eq. (6.6) reduces to Eq. (3.42) as required, since a small value of β corresponds to a situation in which the beam ions constitute only a small fraction of the total ion density, in which case the beam plasma reduces to a normal plasma. A value of β close to unity corresponds to a situation in which almost all of the ions are beam ions. Figure 6.3 shows a number of solutions of Eq. (6.6) with β as a specified parameter.

Each of these solutions has the same general character as the $\beta = 0$ case, the ordinary plasma case discussed in Sec. 3.5, but with values of η_0 and ξ_0 dependent on β. Since

$$(d\xi/d\eta)_{\eta_0} = 0 \tag{6.7}$$

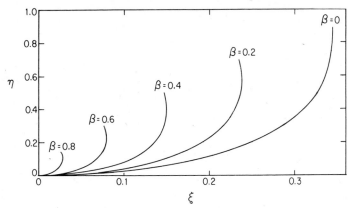

Figure 6.3. Normalized potential $\eta = -eV/kT$ as a function of the dimensionless distance $\xi = x/L$ measured from the median plane of a uniform slab ion beam with a uniform generation rate g of slow ions. $L = (n_0/g)\sqrt{2\,kT/M}$ and the parameter β = beam ion density divided by total ion density.

we can set the derivative of the right hand side of Eq. (6.6) equal to zero at $\eta \approx \eta_0$. This yields

$$[(1 - \beta)/\sqrt{\eta_0}] - 2F(\sqrt{\eta_0}) = 0,$$

which may be written as

$$\beta = 1 - 2\sqrt{\eta_0}\,F(\sqrt{\eta_0}). \tag{6.8}$$

This is plotted in Fig. 6.4 for values of η_0 leading to values of β in the range $0 < \beta < 1$. The corresponding values of ξ_0, obtained by substituting Eq. (6.8) into Eq. (6.6), are given by

$$\xi_0 = [(1 + 2\eta_0)\,F(\sqrt{\eta_0}) - \sqrt{\eta_0}]2/\pi \tag{6.9}$$

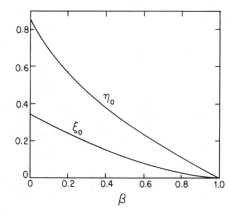

Figure 6.4. Limiting values of η and ξ for the ion beam plasma as a function of β.

and are also shown in Fig. 6.4. As in the case of plasmas with no ion beam the exact solutions must depart from Eq. (6.6), or from the curves of Fig. 3.3 as $\xi \to \xi_0$ and make a transition to the sheath solution as shown schematically in Fig. 3.6.

The variable ξ is distance from the median plane normalized to the length L given by Eq. (6.3). If the beam width matches the channel width, and the sheath thickness is assumed narrow compared to the beam width, then the turnaround value of ξ_0 can be identified with the half width a through the equation

$$\xi \approx a/L \tag{6.10}$$

so that $\xi/\xi_0 \approx x/a$. In Fig. 6.5 η is plotted as a function of this variable.

Usually, the channel boundary is not at the beam edge. In that case the identification of ξ_0 with a/L, where a is the beam width, might be reasonable for β close to 1, but for $\beta \ll 1$ the plasma would extend to the channel walls.

To consider the nature of the sheath solution it is necessary to consider the ion current which drifts to the boundaries and the potential of the beam relative to its boundaries.

A. Slow Ions to the Beam Boundary

Slow ions can be formed within the beam by various mechanisms: ionization of the neutral gas by the fast ions is likely only for very high energy ion beams, ionization by those of the neutralizing electrons which have sufficient energy must also occur, but the most prolific source of slow ions is usually charge transfer between the beam ions and the ambient gas. Even when the ionization is by electrons most of this will occur within the beam. We therefore consider an ion generation rate g which is uniform up to the beam boundary. The resulting current density of slow ions moving transverse to the beam is

$$J_i = e \int_0^a g \, dx = ega. \tag{6.11}$$

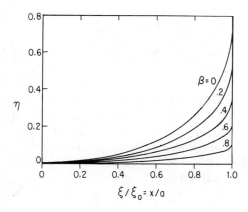

Figure 6.5. Potential variations across the beam plasma as a function of distance normalized to the beam width for the case of a beam width exactly equal to the channel width.

If β is close to 1 we may use Eq. (6.10) even when the walls are far from the beam boundary, obtaining

$$J_i = egL\xi_0. \tag{6.12}$$

But L is given by Eq. (6.3), leading to

$$J_i = en_0\xi_0 \sqrt{2\,kT/M}. \tag{6.13}$$

The degree to which $\beta \approx 1$ is discussed in Sec. 6.2B.

B. Calculation of Beam Plasma Potential

Another important parameter is the potential at the center of the beam relative to the channel boundaries. The required potential is that which ensures that the slow ion loss current is equal to the slow ion generation rate and the electron loss rate matches the electron generation rate. We consider the case in which electrons are introduced peripherally to the beam, as would be the case if they originated as secondaries at the accel electrode and collector, in which no ionization occurs in the beam and in which charge exchange between the fast ions and ambient gas occurs.

Start by equating the ion loss rate given by Eq. (6.13) from each side of the beam to the ion creation rate, obtaining

$$2\xi_0 n_0 e \sqrt{2kT/M} = n_a \sigma_c I_{bl}, \tag{6.14}$$

where n_a is the neutral atom or molecule density, σ_c is the charge transfer cross section, and I_{bl} is the ion beam current per unit beam height. Into this equation substitute $n_0 = n_b/\beta$ from Eq. (6.5) and

$$I_{bl} = 2en_b a \sqrt{2\,eV/M}, \tag{6.15}$$

where eV is the energy of a beam ion, obtaining

$$\frac{\beta}{\eta_0} = \frac{1}{\sigma_c a n_a} \sqrt{\frac{kT}{eV}}. \tag{6.16}$$

Since ξ_0 is a function of β, the ratio β/ξ_0 is a function of β. This function is shown in Fig. 6.6 and it is clear that a value of β/ξ_0 leads to a value of β. Before proceeding to the next step, that of equating the electron injection and loss rates, let us see what sort of values of β are obtained from Eq. (6.16) and Fig. 6.6.

Consider a plasma for which $Ve/kT = 1000$, for example, 5000 eV ions and electron temperature of 5 eV, and $a = 10$ cm. Relatively low values of σ_c and n_a

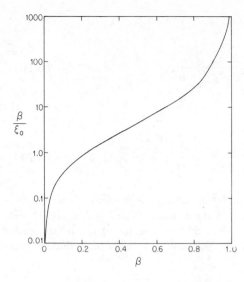

Figure 6.6. The ratio β/ξ_0 as a function of β.

would be 5×10^{-16} cm^2 and 3×10^{11} cm^{-3} ($p \approx 10^{-5}$ torr), respectively. This would lead to $\beta/\xi_0 = 21$ giving a value of $\beta = 0.75$. High values of σ_c and n_a would be 5×10^{-15} cm^2 and 10^{13} cm^{-3} leading to $\beta/\xi = 0.063$ and $\beta \approx 0.02$. To get an ion beam virtually free of slow ions due to charge exchange ($\beta \approx 1$) requires that both charge exchange cross section and background pressure be extraordinarily small. For orientation purposes I note that the cross section for a 5 keV singly charged argon ion with a neutral argon atom is about 10^{-15} cm^2.

We see that slow ions are almost always an important component of the beam plasma and in many cases have a density much larger than that of the beam ions. The small value of β somewhat invalidates the use of Eq. (6.12), a point I shall return to shortly. Here I only wish to point out that the error is in such a direction that the value of β obtained above is too large for small β.

To arrive at a potential of the beam relative to the walls it is necessary to equate the electron generation and loss rates. Let I_{el} be the electron current introduced into the beam per unit of beam height, getting

$$I_{el} = 2ln_0 e \sqrt{kT/2\pi m} \exp - (eV_0/kT). \qquad (6.17)$$

Let $I_{el} = \gamma I_{bl}$ and make the same substitutions which transformed Eq. (6.14) to Eq. (6.15) to obtain

$$\exp \frac{eV_0}{kT} = \frac{l}{2a\beta\gamma} \sqrt{\frac{M}{\pi m} \frac{kT}{eV}}. \qquad (6.18)$$

Combining this with Eq. (6.16) yields

$$\exp \frac{eV_0}{kT} = \frac{l\sigma_c n_a}{2\gamma\xi_0} \sqrt{\frac{M}{\pi m}},$$ (6.19)

leading to

$$V_0 = \frac{kT}{e} \ln \left[\frac{l\sigma_c n_a}{2\gamma\xi_0} \sqrt{\frac{M}{\pi m}} \right].$$ (6.20)

To see the implications of this equation let us see the range of values we might obtain for the bracketed term in Eq. (6.20). Take as fixed value $l = 2$ m and $M/m = 7.3 \times 10^4$, appropriate to argon ions. For this low values of σ_c and n_a used above we obtained $\beta = 0.75$ implying $\xi_0 = 0.04$. With a fairly high value of γ, say $\gamma = 0.5$, we obtain

$$V_0 = 4.7 \, kT/e.$$ (6.21)

The larger values of σ_c and n_a used above led to $\beta \ll 1$ for which $\xi_0 = 0.34$. Take the small value of $\gamma = 0.1$ obtain

$$V_0 = 10 \, kT/e.$$ (6.22)

The point to be observed is the common point that the logarithm of a large number is so insensitive to the number that one can virtually ignore the variation and take

$$V_0 \approx 7 \, kT/e.$$ (6.23)

Crow (1977) considered, in addition to the case discussed here, the case where the fast ions produce ionization in the beam and the case where electrons shot through the beam produce ionization in the beam. We do not feel it significant to cover all cases here. Because the rate of electron escape varies as $\exp(-eV_0/kT)$ all cases lead to a logarithmic variation of potential with values which quite broadly lie between those given by Eqs. (6.21) and (6.22) and adequately approximated by Eq. (6.23).

The insensitivity of the logarithm to the magnitude of the argument renders the question as to whether it was justified to use Eq. (6.12) relatively unimportant. An error by a factor of 3 in the argument only changes the kT/e multiplier in Eq. (6.23) by an additive 1 and even a factor of 10 adds only 2.3. Thus it is possible to say that the potential, in units of the electron temperature, is fairly well determined, even if the lateral ion drift, given by Eq. (6.13), is subject to considerable question.

The calculations of Sec. 6.2A and 6.2B hardly constitute a detailed computation of the potential profile across an ion beam even for the highly idealized one dimensional uniform beam density case considered. It remains to match the solution in the beam to a solution satisfying conditions between the beam and the wall. Then, armed with this information and the electron temperature, the beam divergence could be calculated. As far as I am aware nothing quite so detailed has been carried out, but Crow (1977) did some interesting experiments in which V_0 was measured and correlated with the beam spreading under a variety of conditions. It is of interest to note some of his results.

C.   Experimental Measurements

In Crow's experiments a uniform and unidirectional high poissance beam of 500 eV A^+ ions was produced by using the double fine mesh extraction scheme of Sec. 5.7. The beam spreading defined below was determined by the current to probes which moved across the beam at two positions as indicated in Fig. 6.7. A large planar electron emitter heated by a noninductive element (not shown in Fig. 6.7) could generate a broad electron beam which could be accelerated through the ion beam by a fine tungsten grid. His apparatus also contained an electromagnet as shown, capable of producing a uniform transverse magnetic field over the second half of the beam path. The magnetic field, variable up to about 0.025 tesla, was weak as far as ions were concerned in that the radius of curvature of the beam ions was very large compared to the distance the beam traveled in the field, but strong in that the radius of curvature of thermalized electrons was very small compared to any of the beam dimensions.

The potential of the beam was measured by a determination of the ion velocity in the beam. The velocity was measured by the rate at which a small RF amplitude modulation of the beam propagated, obtained from the phase shifts between two probes in the beam. This measurement was confused by a fast wave propagating in the ion beam plasma but the confusion could be eliminated by measuring the phase shifts between successive maxima in the beam where the two waves were in phase. By this means Crow was able to measure the beam potential to within 5 volts giving, through Eq. (6.23), the electron temperature to less than 1 volt.

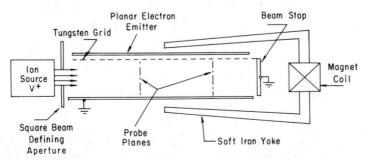

Figure 6.7. Schematic drawing of the experimental setup of Crow.

The primary parameter by which beam spreading was measured is illustrated in Fig. 6.8. In all cases what was found was that a uniform density region existed in front of the source and that this region converged at an angle θ_b, which was the angle used as the criterion for spreading. If the spreading were solely due to the spread in angles θ_f at which the ions were launched, then, as seen in Fig. 6.8a, the two angles θ_b and θ_f would be equal. In these experiments θ_b was always much greater than the estimated value of θ_f, which was only about 0.5°, indicating that the ions were peeled off by sheath effects as shown in Fig. 6.8b.

Under the simplest operating conditions, that is, with no electrons injected across the beam and no magnetic field, the potential of the beam relative to the walls was zero within the 5 volt accuracy of the experiment indicating an electron temperature less than 1 eV. For a given poissance the angle θ_b was approximately constant, although there was a slight increase with voltage which presumably caused a small increase in electron temperature. As poissance went from 2 to 10 to 50 the angle θ_b went from 2.7° to 2.2° to 3.6°. No simple dependence on poissance was evident.

Chamber pressure during operation was normally in the low 10^{-5} torr range. If this was permitted to rise, the beam spreading decreased. This commonly observed effect, usually attributed to improved neutralization (e.g., by Coupland et al., 1973), is in the framework of the analysis of the previous section, due to a cooling of the electrons by collisions with the neutral gas.

The injection of electrons across the beam, normally at energies of about 10 eV, which might have been expected to improve the beam propagation, always had a strong effect in the opposite direction, that is, the angle θ_b was significantly increased. The effect is presumably due to a heating of the plasma electrons and therefore an increase in the beam potential, although it remained zero within the accuracy of the measurement.

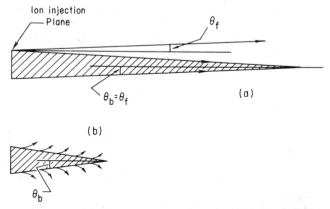

Figure 6.8. Illustration of the region of constant beam intensity (shaded area) for the case (a) where the ions all maintained their launch angles $\theta \leq \theta_f$ and (b) where the beam peeling occurs as a result of sheath penetration into the beam.

The introduction of a transverse magnetic field of the order of 0.01 tesla raised the beam potential to an easily measurable 30 volts, which would imply an electron temperature of ~ 5 eV and the angle θ_b was greatly increased. The magnetic field makes it difficult for the beam to obtain its neutralizing electrons, and for electrons to make their way throughout the beam, from the target at which they are generated, the beam appears to create microinstabilities which enable the electrons to walk across the field. It is surmised that the field fluctuations which permit the electrons to move across the field also produce increases in the electron temperature which cause the beam potential to rise. This in turn increases the penetration of the sheath into the beam and increases the rate at which ions peel off.

Chapter 7

Ion Source Cathodes

7.1 Importance of the Cathode Problem

Ion source discharges for high current ion sources generally require cathodes that must supply hundreds to thousands of amperes. The material and configuration of these cathodes in large measure will determine or at least be linked to virtually all of the important properties of the ion source such as lifetime, stability, turn on time, available pulse length, required source power, available plasma density, and electron temperature. Thus an understanding of cathodes is vital.

7.2 Free Electron Theory of Thermionic Emission

Thermionic emission is the evaporation of electrons from the surface of a solid. Although this subject is well covered in many books (e.g., Seitz, 1940; Kittel, 1966, it seems important to include here at least the rudiments of the elementary theory of the phenomenon. In the free electron picture of a conductor the conduction electrons are considered to exist in a perfectly flat bottomed potential energy well of depth W as shown in Fig. 7.1. Within this well the electrons must obey Fermi–Dirac statistics, that is, the number of electrons per unit volume of phase space (x, y, z, p_x, p_y, p_z) is given by

$$f(\mathcal{E}) = (2/h^3) \left\{ \exp \left[(\mathcal{E} - \mathcal{E}_F)/kT \right] + 1 \right\}^{-1}, \tag{7.1}$$

where h is Planck's constant, $\mathcal{E}_F$ is an energy level called the Fermi energy, and

$$\mathcal{E} = (p_x^2 + p_y^2 + p_z^2)/2m \tag{7.2}$$

155

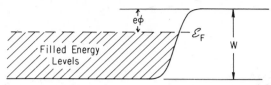

Figure 7.1. Free electron picture of the potential energy of an electron near the surface of a metal.

is the kinetic energy of an electron mass m whose momentum components are p_x, p_y, and p_z. The Fermi energy must be chosen so that the integral over all p_x, p_y, and p_z gives the volume density of the electrons in the conductor.

For the electron densities characteristic of metals and other good conductors $\mathcal{E}_F \gg kT$ right up to the melting point of the material. In that case the Fermi function (7.1) has the characteristic shape given by Fig. 7.2: it is substantially flat until the electron energy comes to within, say, $3kT$ of $\mathcal{E}_F$ and then drops rapidly passing through $f(\mathcal{E}) = 1/2$ at $\mathcal{E} = \mathcal{E}_F$. For $(\mathcal{E} - \mathcal{E}_F)/kT \gtrsim 3$ the decay is purely exponential, as in the classical Boltzmann distribution. We speak, roughly, of the conduction band being filled up to the energy $\mathcal{E}_F$ and empty for electron energies $> \mathcal{E}_F$, although this description is strictly true only in the limit as $T \rightarrow 0$ K. The energy which must be added to an electron at the top of the filled levels, that is, an electron for which $\mathcal{E} = \mathcal{E}_F$, to give it an energy equivalent to an electron at rest outside the metal is designated as $e\phi$. The quantity ϕ is termed the work function.

If one takes the escape surface as a plane of constant z, then it is only necessary to multiply Eq. (7.1) by $v_z = p_z/m$ to obtain the rate at which particles per unit area per second per unit volume of momentum space strike the surface. Assuming that no reflection occurs for those electrons with sufficient z directed velocity to escape, we obtain for the current density

$$J = (2e/mh^3) \int_{-\infty}^{\infty} \int_{-\infty}^{\infty} \int_{p_0}^{\infty} p_z \left\{ 1 + \exp\left[\frac{(\mathcal{E} - \mathcal{E}_F)}{kT}\right] \right\}^{-1} dp_x dp_y dp_z, \quad (7.3)$$

where $p_0 = \sqrt{2mW}$. With $e\phi/kT$ taken $\gg 1$, Eq. (7.3) can be shown to lead to the Richardson equation

$$J = AT^2 \exp(-e\phi/kT), \quad (7.4)$$

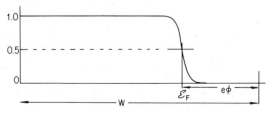

Figure 7.2. Fermi–Dirac function $\{1 + \exp[\mathcal{E} - \mathcal{E}_F)/kT\}^{-1}$ for electrons in a conduction band.

where

$$A = 4\pi mek^2/h^3 \qquad (7.5)$$

is a universal constant of value 1.2×10^6 A/m²K² or 120 A/cm²K².

To verify the nature of Eq. (7.4) and determine the constants A and ϕ, thermionic emission current density data are obtained over as large a range as possible and plotted as in Fig. 7.3. The straightness of the line obtained tends to verify the validity of the Richardson–Dushman equation and the slope of the line yields the thermionic work function ϕ. However, the value of A obtained by extrapolating to $1/T = 0$ does not ordinarily turn out to be 120 A/cm²K². A few values are given in Table 7.1.

Although values of A vary considerably from the theoretical value a substantial degree of correctness of the treatment is indicated by the fact that for pure metals the values lie within a factor of 0.25–1.33 of the theoretical value and depart by very large numbers only for composite surfaces where the simple assumptions about the surface can hardly be expected to be valid. Even without critical evaluation it seems worthwhile to mention some of the approximations which were made in the treatment leading to Eqs. (7.4) and (7.5), as follows:

1. The free electron picture in which the interaction with the lattice and all other electrons is approximated by a flat potential.

2. A related assumption that the equivalent electron mass inside the metal is identical with the mass of a free electron.

3. The treatment of the surface as planar even on an atomic scale.

4. The assumption of zero reflection coefficient for all electrons with a velocity adequate to escape the metal.

5. Possible distortion of the electron distribution by the escaping electrons.

It does not appear surprising that the constant A shows so much variation, and beyond those approximations another source of error is implicit in the way the quantity is measured. The depth of the filled levels depends on the electron density.

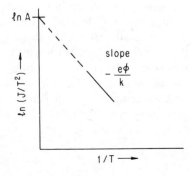

Figure 7.3. Richardson plot for the determination of the constants ϕ and A.

TABLE 7.1   Representative Thermionic Emission Data

Metal	$A(\text{A}/\text{cm}^2\text{-K}^2)$	$\phi(\text{volts})$
W	70	4.5
Ta	55	4.2
Ni	30	4.6
Cs	160	1.8
Pt	32	5.3
Cr	48	4.6
Ba on W	1.5	1.56
Cs on W	3.2	1.36

Thus the thermal expansion must lead to a work function ϕ varying with temperature. If we write

$$\phi = \phi_0 + \alpha T \tag{7.6}$$

where ϕ_0 is the work function at absolute zero, then Eq. (7.4) becomes

$$J = AT^2 \exp\left(-e\alpha/k\right) \exp\left(-e\phi_0/kT\right). \tag{7.7}$$

leading to a measured value of A different from that given in Eq. (7.5) by the factor $\exp\left(-e\alpha/k\right)$. This factor alone may go far to explain the disparity.

7.3   The Schottky Effect

Curves of the type shown in Fig. 7.3, from which the data given in Table 7.1 are obtained, are not as easily plotted as might be imagined. In one set out to measure the current density J from an electron emitter as a function of the temperature T, with, say, a diode containing a cylindrical cathode coaxial with a larger radius cylindrical anode, one might naively expect curves such as those seen in Fig. 7.4 giving the emission limited current density J as a function of T. Instead, experimental curves show shapes like those in Fig. 7.5. The small variation in current among the curves for different temperatures in the space charge limited region is easily understood in terms of the maxwellian distribution of the emitted electrons, as covered in Sec. 2.9. If $V \gg kT/e$, this is small and presents no difficulty. The difficulty in obtaining a well defined value of J corresponding to a given T is that the emission current shows no saturation and breaks away from the space charge limited curve in a gradual fashion. This effect is due to the lowering of the work function by the external field.

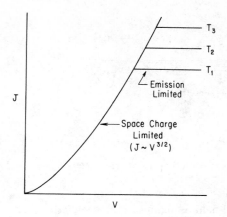

Figure 7.4. Idealized vacuum diode characteristics for various temperatures of the emitter.

As an electron escapes from the surface of a conductor it is ultimately restrained by an image force

$$F_i = -e^2/(16\pi\varepsilon_0 z^2),$$ (7.8)

where z is distance from the plane boundary in which its charge is imaged. This force is obtainable from a potential

$$V_i = -e^2/(16\pi\varepsilon_0 z).$$ (7.9)

A potential which has the same character as Eq. (7.9) at large z but leads to a finite well depth W at $z = 0$ is given by

$$V = -e^2/(16\pi\varepsilon_0 z + e^2/W),$$ (7.10)

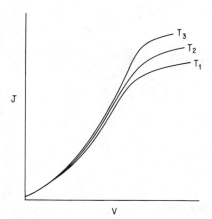

Figure 7.5. Realistic vacuum diode characteristics.

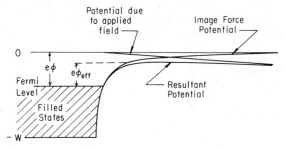

Figure 7.6. Illustration of the lowering of the work function by an external accelerating field.

as illustrated in Fig. 7.6. When an external field of magnitude E accelerating electrons away from the surface is applied the resultant potential is

$$V = -Eez - e^2/(16\varepsilon_0 z + e^2/W).$$ (7.11)

As seen in Fig. 7.6 potential goes through a maximum for which the effective work function, ϕ_{eff}, is lower than the zero field work function. The calculation of the amount of lowering leads to a modified expression for the current density,

$$J = AT^2 \exp\left(-e\phi/kT\right) \exp\left[(e/kT)\sqrt{Ee/4\pi\varepsilon_0}\right].$$ (7.12)

This effect was first analyzed by Schottky (1914) and is known as the Schottky effect.

Equation (7.12) is verifiable by plotting $\ln J$ as a function of $\sqrt{E}$, as shown in Fig. 7.7, where E is the field intensity at the cathode surface calculated from the geometry and applied potential. The E calculated in this manner is the actual elec-

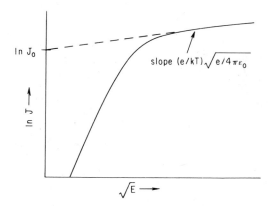

Figure 7.7. Verification of the Schottky effect and determination of the zero field thermionic emission current density.

tric field at the cathode surface only for potentials substantially larger than those in the space charge limited region.

All pure metals do exhibit quantitatively the shape shown in Fig. 7.7 including the transition to the calculated Schottky slope. This not only constitutes a verification of the theory but provides the means of extrapolating the emission to the zero field value for the purpose of making Richardson plots, such as shown in Fig. 7.3, to obtain the constant ϕ and A characterizing the emitting surface.

For composite emitters such as cesium or barium on tungsten, various oxides, and dirty surfaces generally, the emitting surface is made up of a patchwork of elements at different potentials. These local potential variations produce fields which retard electron escape over much of the surface. In this case very large field effects on emission are observed since increasing potentials increase the effective emitting area. Such a large field effect is called the anomalous Schottky effect. Ultimately, at fields large enough to be accelerating over all portions of the emitter, the curve of ln J versus $\sqrt{E}$ approaches the normal Schottky slope even in those cases.

The effect of electric fields on electron emission, which is called the Schottky effect, should not be confused with the phenomenon known as field emission. When fields become of the order of 10^8 V/cm, fields realistically achievable only at sharp points, then the barrier for electron escape becomes so narrow that it no longer becomes necessary for electrons to surmount the barrier. Electrons can escape the metal by tunneling through the barrier. This is not a thermionic effect and can occur at arbitrarily low temperatures.

7.4   Power Balance at a Cathode

The delivery of power to a cathode often is done by direct ohmic heating; in other cases it is done by radiation from an enclosed heater. The cathode loses power by radiation and conduction. In addition, there are two other important sources of heating and cooling. Each electron which evaporates from a source carries off an energy $e\phi$ so that an emission current density J_e results in cooling power $J_e\phi$ per unit area. In a plasma ions strike the cathode with an energy eV, where V is the voltage between the cathode and the plasma, so that there is an additional heating power density J_iV. It is interesting to analyze the relative magnitudes of the electron cooling and ion heating terms when the emission from the cathode is space charge limited. The ratio of the currents, given by Eq. (3.84), leads immediately to

$$J_e\phi/J_iV = \sqrt{M/m}(1 - \sqrt{2kT/eV})\phi/V. \qquad (7.13)$$

Suppose, for example, we take $\phi = 4.5$ volts, as for tungsten, $\sqrt{M/m} = 43$, as for H^+ ion, $kT/e = 5$ and $V = 80$ volts, values that might readily be obtained for a hydrogen discharge. This would yield a power lost which was twice the power delivered by ions. In the space charge emission limit the cathode would be cooled by the combined processes of ion bombardment and electron emission. For most realistic cases $J_e\phi/J_iV$ would be even larger. This is a stabilizing effect. The ion beam heats the cathode but cannot carry it up to the point of space charge limited emission before cooling effects stabilize the cathode temperature.

For intense discharges, however, the ion current, given by Eq. (3.28), may be so great that the electron current which can be calculated from Eq. (3.94) is beyond the capability of the cathode. In such cases it is possible to destroy the cathode by the power delivered by the bombarding ions, either by melting or evaporating the surface.

In some sources the cathode heater is turned off after the discharge starts. In others there is no cathode heater with the discharge being started with high voltage or radio frequency. In these cases the cathode reaches a temperature such that the heating by ions is exactly balanced by the cooling due to radiation and electron emission. It can easily be shown using Eq. (7.13) that such operation can never have space charge limited electron emission.

With a heater, space charge limited emission can be reached but it would be a very unusual mode of operation. It would result, for example, in a situation where the cathode would get hotter and perhaps burn out if the gas flow or arc voltage were turned off.

The conclusion reached in this section that space charge limited emission would be unusual contradicts a common belief. One can turn to virtually any discussion of discharge phenomena to see references to space charge limited operation (e.g., Wilson and Brewer, 1973, p. 50; Goede and Green, 1982, p. 1800). The state of a discharge referred to in this way is one in which the electron evaporation cooling and ion bombardment heating are both important terms in the cathode power balance. In that case the current will increase with any change in the discharge that increases the ion bombardment power, such as increasing the arc voltage or increasing the plasma density. However, the emission will rarely be space charge limited in the sense of having a potential minimum in the cathode sheath or zero potential gradient at the cathode surface.

7.5   Tungsten, Tantalum, and Molybdenum

In the simplest type of electron source, emission is achieved by passing current through the uncoated refractory metal until the wire is ohmically heated to a sufficiently high temperature. The three metals that may be considered for this purpose are tungsten, tantalum, and molybdenum. The zero field emission currents for these three metals as a function of temperature are shown in Fig. 7.8. The temperature at which the emitter can operate is ultimately limited by the melting point of the material, but even at much lower temperatures the rate of evaporation limits the operating temperature. It is therefore useful to plot the evaporation rates as a function of the current density for these three metals.

Vapor pressure data of these materials are readily found tabulated in handbooks.

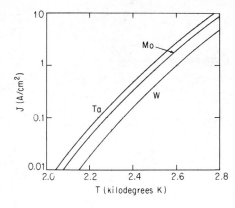

Figure 7.8. Thermionic emission current density from molybdenum, tantalum, and tungsten as a function of temperature.

Kinetic theory gives the rate of evaporation in atoms per unit area per unit in terms of the pressure and temperature as

$$\nu = p/\sqrt{2\pi MkT}.$$

The mass per unit and per second can be obtained by multiplying by the atomic mass M. If we divide the result by the density, we obtain the evaporated thickness per second,

$$S = (p/\rho)\sqrt{M/2\pi kT}. \tag{7.14}$$

Equation (7.14) must be in a consistent set of units such as mks units. With S in centimeters per second, p in torr, ρ in grams per cubic centimeter, and T in kelvins we obtain

$$S = 5.87 \times 10^{-2}(p/\rho)\sqrt{M_A/T}, \tag{7.15}$$

where M_A is the mass in atomic mass units. If we plot S as a function of J for the three materials we have been considering, we obtain Fig. 7.9. It is clear that tantalum and tungsten make comparable emitters by this criterion, but that molybdenum is not a competitive emitter material. In Fig. 7.9 it appears that tungsten is somewhat better than tantalum at high current densities. However, the difference is probably within the limits of error with which the various parameters that went into the calculation are known. Other criteria—stiffness at high temperature, for example—tend to favor tungsten, so that tungsten is usually, among the pure metals, the emitter of choice.

   If we take 10^{-3} cm as a significant thinning, we obtain, at 10 A/cm^2, a time to such thinning of about 10 hours. This may seem a conservative point of operation as long as the filament radius $\gg 10^{-3}$ cm. Actually, nonuniformities in temperature will cause the hotter areas to thin faster, lead to further overheating,

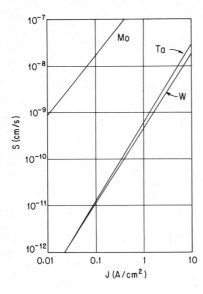

Figure 7.9. Surface evaporation rate as a function of thermionic emission current density for molybdenum, tantalum, and tungsten.

and cause a runaway failure. In practice, for long term stable operation 1 A/cm^2 is regarded as a maximum for tungsten, as it is for most good electron emitters.

7.6   Some Thin Film and Oxide Emitters

The subject of electron emitters is broad enough that books (e.g., Jenkins and Trodden, 1965 and voluminous papers such as that by Nottingham (1956) and Kohl (1967, Chap. 16) are devoted entirely to this topic. Nevertheless, it seems necessary to survey here many types of emitters in order to point out which are suited for use in ion source generators.

Good electron emission can be obtained from a refractory metal coated with some fraction, usually about 2/3, of a monolayer of certain other elements or compounds. A common emitting surface of this kind is the thoriated tungsten emitter. In this emitter, thoria, ThO, is incorporated in the tungsten during manufacture. When the tungsten is heated to 2500 K for 30–60 seconds some of the thoria is reduced to metallic thorium. When the tungsten is held at 2100 K some of the thorium diffuses to the surface. The tungsten may then be operated at a temperature of 1800–2200 K, yielding a peak emission of approximately 3 A/cm^2.

The lifetime of a thoriated tungsten emitter is greatly improved by a carburization of the surface. This is accomplished by subjecting the formed and mounted filaments to a heat treatment at about 2200 K in a mixture of hydrogen and some hydrocarbon such as benzene or acetylene. The tungsten carbide, W_2C, formed at the surface diffuses into the surface for a small distance.

The carburized thoriated tungsten emitter is a useful cathode material but is not very favorable as a plasma cathode. In a plasma the bombarding ions too rapidly deplete the surface thorium. There are similarities with the lanthanum oxide doped molybdenum (LM) cathode discussed in Sec. 7.10, which does make a good plasma cathode, but we shall come to that later.

The cesiated tungsten surface, examined in detail in Sec. 9.4., is another example of a thin film emitter. In this case cesium must be delivered continuously to the surface as it evaporates. Such a renewable cathode, especially in a hollow cathode configuration, as discussed in Sec. 7.12, is an ideal cathode for a cesium discharge, a somewhat limited application.

A widely used electron emitting surface is created by coating a nickel surface with a mixture of barium, calcium, and strontium carbonate. When these materials are heated in a vacuum CO_2 is evolved, leaving behind a mixture of the oxides BaO, CaO, and SrO. This mixture makes an excellent emitter with current densities of the order of 3 A/cm^2 available at temperatures as low as 850°C. Since the material is laid down in some thickness the cathode can stand considerable sputtering and therefore can be used in plasma devices such as ion source discharges. It has the disadvantage of sensitivity to impurities, especially water vapor. With some care to be sure that the cathode is warm and exposed only to dry gases, an oxide cathode, after activation, can be taken from vacuum to atmospheric pressure and still function when brought back to vacuum conditions. Usually, however, such cathodes must be stripped, recoated, and reactivated after each exposure.

7.7 The Hull Dispenser Cathode

A very good plasma cathode due to Hull (1939) has had very little application to ion source work in spite of the fact that it is well suited to the purpose. In this cathode a closely woven cylindrical mesh of fine molybdenum wire (similar to the braided copper often used as shielding for electric cables) is filled with small granules of fused baria-alumina eutectic (70% Ba, 30% Al_2O_3 by weight). This filled flexible porous molybdenum tube placed inside a heat shielded structure containing molybdenum surfaces, such as that shown in Fig. 7.10, serves both as the cathode heater and as the dispenser of active material with coats the molybdenum vanes. When operated at a temperature of 1150–1200°C it dispenses the active material, barium oxide, at just the right rate and at the same time maintains the emitting surfaces at a temperature of the order of 800°C where copious emission can be obtained. For example, the cathode shown in Fig. 7.10, operated at 831°C, yielded an emission current of 360 A for a current density of 1.8 A/cm^2. Another smaller cathode yielded 64 A for a current density of 3.2 A/cm^2 at 766°C. Neither of these numbers is quite consistent with the emission constants of the material measured by Hull in a special tube for the purpose. The values obtained were

$$A = 0.85 \ A/cm^2K^2$$

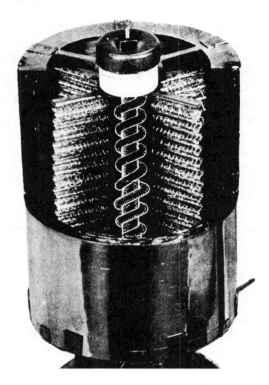

Figure 7.10. Dispenser cathode of A. W. Hull. The central spiral is a braided cylinder of molybdenum which contains a baria-alumina eutectic. This heats the cathode and continuously dispenses barium oxide to the emitting molybdenum surfaces. From Hull, 1939.

and

$$\phi = 1.215 \text{ volts.}$$

Hull did not report on the maximum available current density from these cathodes. In a geometry in which BaO from the emitting surfaces did not readily find the cathode aperture but bounced from surface to surface, it is likely that higher operating temperatures and therefore higher current densities could have been obtained along with lifetimes adequate for most purposes. Hull's lifetimes were very long. Several cathodes were run in mercury discharges for 20,000 hours and one for 30,000 hours without diminution in the cathode performance.

The baria-alumina eutectic material is stable to air and the vanes have such a thin coating of BaO that there is a possibility that this cathode could be let down to air, after running, and then restarted. If so, that would be a significant advantage beyond those investigated by Hull.

It is not clear why this cathode has not had a wider application to plasma devices including ion source plasmas. I surmise this is so because it was never developed into a simple commercial package. A researcher would have to obtain the molybdenum stocking material and the baria-alumina eutectic material and construct a cathode appropriate to his needs. This is in contrast to the cathodes described in the next section, which are cathodes one does not characteristically prepare but rather purchases in neat packages. For most developmental ion source work the ability to construct cathodes to suit one's needs has a substantial advantage over being restricted to prepackaged shapes for which there is a large market.

7.8 Dispenser, Impregnated, and Pressed Cathodes

Although the distinction is not usually made, I find a substantial difference between the Hull-type dispenser cathode, in which active material is evaporated from a heater to another surface, and those cathodes to which the term dispenser cathodes commonly refers, including the impregnated and pressed types, in which the active material percolates through the structure itself to the emitting surface.

The first of these was the L-type cathode illustrated in Fig. 7.11. In the configuration illustrated, a pellet of a mixture of the carbonates is placed in a well in a molybdenum cylinder which is capped, as shown, by a porous tungsten disk. The carbonates are reduced to the oxides by heating and then the cathode operates through a slow flow of the oxides through the porous tungsten to the emitting surface. The activation temperature for this cathode is extremely critical; the activation process takes about 2 hours followed by an aging treatment extended over several days.

These problems were overcome by replacing the Ba–Sr carbonate pellet with a pellet of a mixture of tungsten powder and a Ba–Ca aluminate. Levi (1953, 1955) simplified the structure and improved the performance by directly impregnating the porous tungsten with the barium–calcium aluminate as shown in Fig. 7.12.

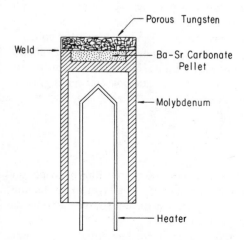

Figure 7.11. L cathode.

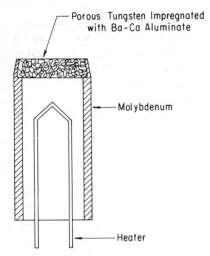

Porous Tungsten Impregnated
with Ba-Ca Aluminate

Molybdenum

Heater

Figure 7.12. Impregnated cathode.

This impregnated cathode has produced current densities of 9 A/cm^2 at 1130°C under pulsed conditions.

Another type of dispenser cathode is the pressed cathode. In this case a button is made by pressing a disk of a mixture of a powder of a Mo–W alloy and powdered Ba–Ca aluminate and then sintering in vacuum or hydrogen to 1750°C. This forms a hard strong button that replaces the impregnated porous tungsten disk in Fig. 7.12.

The dispenser cathodes described in this section poison readily. For example, oxygen exerts a poisoning effect at partial pressure levels as low as 10^{-7} torr. Hydrogen exerts a poisoning effect on Ba–Ca aluminate cathodes for pressures greater than 5 × 10^{-6} torr. Metal vapors, with the exception of copper vapor, have a harmful effect and these cathodes cannot be operated in mercury vapor.

Another type of pressed and sintered cathode which has been studied in depth uses nickel rather than tungsten. In the ''mush cathodes'' nickel powder was sintered to a nickel base and carbonates then brushed on. In the ''BN,'' for bariated nickel, cathodes nickel powder is mixed with the carbonates and a reducing agent, pressed into pellets, and sintered. I am unaware of any significant advantages of these nickel based cathodes over the tungsten based cathodes and having mentioned them press on to other types of emitters.

7.9 Lanthanum Hexaboride

Although lanthanum hexaboride might be classified among thin film or dispenser cathodes, insofar as its operation depends on the delivery of a layer of lanthanum to the surface of the boride lattice, it is of such interest for ion source cathodes as to warrant a separate section.

The rare earths, the alkaline earths, and thorium form interstitial compounds

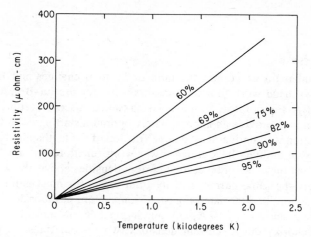

Figure 7.13. Resistivity of LaB$_6$ as a function of temperature for various fractions of maximum density. From Leung, 1987.

with boron. In these hexaborides the boron matrix forms cages which trap the metal atoms. It appears that there are no valence bands between the boron and the trapped atoms leaving the outer electrons of the latter free to form a conduction band. These materials thus acquire a metallic conductivity. High density LaB$_6$, for example, has a resistivity of 13 μohm-cm at room temperature, a value not very different from that of iron or tantalum, for example. Like a metal, the resistivity rises with temperature, as shown in Fig. 7.13. The melting points of the hexaborides are all about 2200 K with that of LaB$_6$ being 2210 K. Because LaB$_6$ has the most favorable emitting properties of the hexaborides I confine my attention to it.

Because the material is commercially available* as rods over a wide range of diameters, I need say little about the material preparation. This is described in some detail by Lafferty (1951), who first gave the detailed emission properties of this material. The processing consists first in the preparation of a pure powder and then pressing and sintering to a dense, purple, very hard ceramiclike material, which can be shaped subsequently only by a difficult grinding operation.

When this material is heated to a sufficiently high temperature, the lanthanum atoms at the surface evaporate and are replaced by atoms diffusing from below. The boron framework remains intact. The diffusion of lanthanum to the surface, as vacancies appear, is what maintains an active electron emitting surface.

Several others, besides Lafferty, have studied the electron emitting properties of LaB$_6$ including, for example, Jacobson and Stroms (1978) and Pelletier and Pomot (1979). The constants for the Richardson equation found by Pelletier and Pomot are

$$A = 120 \text{ A}/\text{cm}^2\text{K}^2$$

*E.g., CERAC, Inc., Milwaukee, Wisconsin.

and

$$\phi = 2.36 \text{ eV}$$

As the lanthanum evaporates the ratio of La to B changes and Jacobsen and Storms (1978) made work function measurements for the La–B system between LaB_4 and LaB_{29}. For purposes of most plasma cathode users it is sufficient to know that LaB_6 seems to provide an approximately optimum starting point for a cathode that will function well until at least a large fraction of the lanthanum is evaporated.

Barium oxide emitters have much lower work functions but are so limited in temperature that they cannot be used at current densities above about 1–3 A/cm^2. Approximately the same current density limit exists for most emitters including tungsten, tantalum, and thoriated tungsten. Lanthanum hexaboride can readily be taken to current densities of 10 A/cm^2 and in practice, presumably because of the anomalous Schottky effect, current densities two to four times as great can be obtained. Goebel et al. (1985) report the emission of 600 A from a flat circular disk for an average of 20 A/cm^2 continuously for 400 hours with no decrease in performance. In Fig. 7.14 a comparison between the emission from W and LaB_6 as a function of temperature is displayed.

Lafferty measured the rate of evaporation of La from LaB_6, and using the density 8.2×10^{21} La atoms$/cm^3$, also from Lafferty, it is possible to obtain the depth of La depletion as a function of temperature. Figure 7.15 is a plot of depletion depth rate as a function of current density, together with a curve of surface evaporation rate of tungsten. This comparison only begins to disclose the advantage of the LaB_6 relative to W. The high temperature required for W implies direct ohmic heating of a filament, which means that very little evaporation can be tolerated before the filament will develop local hot areas and destruct. The LaB_6 on the other hand cannot as readily be made into filaments and is normally radiantly heated. The LaB_6 can be thick and run until all of the La is depleted. The allowed rate of loss of material is much greater than for a tungsten filament.

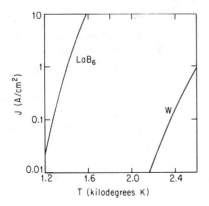

Figure 7.14. Zero field emission current densities from LaB_6 and W.

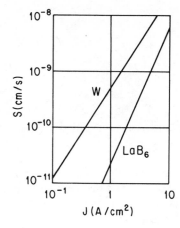

Figure 7.15. Evaporation rate of La from LaB₆ compared with the evaporation rate of W, as a function of current density.

Although LaB₆ is not as readily made into filaments as tungsten, Pincosy and Leung (1984) showed that such filaments have substantial advantages relative to tungsten filaments. In particular their lifetime, for an equivalent emission current, is much longer than the lifetime of tungsten filaments.

In plotting the LaB₆ curve of Fig. 7.15 the starting point was Lafferty's curve of mass evaporation rate as a function of current density. Pelletier and Pomot (1979) found that a layer of Mo on the LaB₆ degrades the performance and it appears that Lafferty's surfaces were slightly poisoned. If the poisoning agent reduced the emission, but not the evaporation rate, then the LaB₆ curve in Fig. 7.15 should be moved substantially to the right so that the contrast between that material and tungsten is even greater than displayed in the figure.

In general one of the great advantages of LaB₆ is its freedom from poisoning effects. Such gases and vapors as O_2, H_2O, and Cl_2 which are devastating to most active cathode materials, are tolerated by LaB₆. The material can even be subjected to an abrupt vacuum failure while hot and start right up again.

If the material is to be run at low current densities (low temperatures), it would have to be activated by raising to an elevated temperature. For current densities of $1 \ A/cm^2$ or higher the material can simply be taken to the emitting temperature and run.

One difficulty in the use of LaB₆ lies in the fact that at elevated temperatures the material cannot be in contact with any of the common refractory materials. In such contact boron invades the refractory metal producing a very brittle, even crumbly structure. The solution to this problem is to clamp the LaB₆ in graphite, which poses no great difficulties. Contact with rhenium or carburized tantalum is also satisfactory.

The material is subject to cracking under thermal shock. In the cathode shown in Fig. 7.16, for example, which was used in my laboratory with considerable success, the LaB₆ was in the form of a capped cylinder. Under a sudden thermal load a fraction would cause the end cap to break away from the cylindrical portion

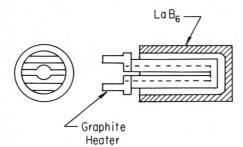

LaB$_6$

Graphite
Heater

Figure 7.16. Example of a LaB$_6$ cathode and heater (to scale).

of the LaB$_6$ emitter. If I were to redesign that cathode, the end cap would be a separate piece held against the cylinder by the same graphite part that supports and provides electrical conduction to the LaB$_6$. Goebel et al. (1985) reported that their 7.1 cm diameter flat disk also developed radial cracks but managed to stay together. It appears that the cracking of large pieces of LaB$_6$ is hard to avoid, and wherever possible, a large area cathode should be made in several interlocking pieces to allow for the effects of thermally generated stress.

Although this book is not oriented toward mechanical details, I call attention to the heater in Fig. 7.16. Tungsten might seem to be the natural heater material, since it can so readily be formed into coils, but the sagging of such heaters proved to be a common source of failure. The graphite heater shown proved to be an excellent solution to that problem and is very easily machined and clamped.

Among several useful directly heated LaB$_6$ cathode designs developed at Lawrence Berkeley Laboratory, that are described by Leung, Moussa, and Wilde (1986) and shown in Fig. 7.17 is particularly worthy of description. The LaB$_6$ emitter is a hollow cylinder 11 mm in diameter and 35 mm long with caps on the ends as shown. (The prospective user should be warned that this is a difficult shape to grind). The assembly is held under tension by a spring on the mounting shaft which pulls the center of the assembly to the left and pushes the outer part of the line to the right of Fig. 7.17. Aside from the compactness of this assembly the fact that there is no magnetic field at the emitting surface on the LaB$_6$ due to the heater current is an important asset for some applications. The field which normally exists at the surface of directly heated cathodes makes the escape of very low energy electrons difficult. This cathode was designed for the purpose of introducing electrons into a plasma with a very small potential difference between the emitter and plasma.

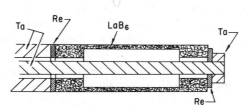

Ta   Re   LaB$_6$ Ta

Re

Figure 7.17. Direct heated concentric LaB$_6$ cathode.

7.10   Lanthanum Oxide Doped Molybdenum Cathodes

The so called LM cathode, actually lanthanum oxide (La_2O_3) doped molybdenum, was first described in a patent by Buxbaum and Gessinger (1977) and in a technical article by Buxbaum (1979). The material is made by mixing 2% by weight of La_2O_3 with Mo powder and then pressing and sintering into desired shapes. This material has mechanical properties very close to those of pure molybdenum so that it may readily be machined and bent into desired shapes.

As in the case of thoriated tungsten, performance and lifetime are greatly improved by carburizing the surface. Emission was also enhanced by the inclusion of about 0.5% of platinum in the powder mix. This cathode has been used in various plasma devices by Goebel et al. (1980, 1985) and Schechter and Tsai (1981).

Although excellent performance has been obtained for the LM cathode in intense discharges, with emission current densities of close to 10 A/cm^2 it has several disadvantages relative to LaB_6 as a cathode material. Whereas the LaB_6 requires no conditioning and shows consistent operation from piece to piece, the LM cathodes must be carefully fabricated and conditioned for good operation and may require about 10 hours of conditioning for good operation. Good operation corresponds to only one half the current density obtainable from the LaB_6.

To sum up this comparison LaB_6 is capable of the higher current density and more reproducible performance but is difficult to shape and is subject to thermal fracture. If it can be used in simple stable shapes, it may be the superior material. The LM material is, at best, capable of yielding only about half the current density of LaB_6 and requires considerable care in preparation and conditioning to achieve reproducibility. Nevertheless, it is capable of being used in configurations which either cannot be formed with LaB_6 or would fracture if it were. Each of these will have its place as ion source cathode materials.

7.11   Liquid Mercury Cathodes

It has long been known, as described, for example, by Cobine (1941, Chap. 11), that a mercury pool made an excellent cathode for a mercury discharge. An emitting bright spot estimated by Prince (1972) to be of the order of 2.5×10^{14} cm^2/A is observed to be in continual motion over the surface of the mercury at a speed of about 10 m/s due to forces exerted by the bombarding ions and the reaction of the vapor streaming from the spot. Such spot motion can be prevented by a piece of tungsten or molybdenum on the surface of the mercury.

King et al. (1967) made the liquid mercury* into a practical cathode for Hg^+ ion sources by force feeding of the mercury through a small orifice in a molybdenum structure as illustrated in Fig. 7.18. Because the heating of the cathode by

---

*Although King et al. refer to their cathode as an LM cathode, that designation must now be preserved for the La_2O_3 impregnated Mo cathode described in Sec. 7.10.

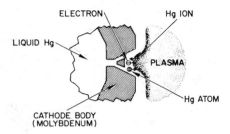

ELECTRON   Hg ION

LIQUID Hg

PLASMA

Hg ATOM

CATHODE BODY
(MOLYBDENUM)

Figure 7.18. Force fed liquid mercury cathode of King, et al. From Brewer, 1970. p. 141.

ion bombardment can easily cause an excessive evaporation of mercury it is necessary to provide cooling for this type of cathode.

7.12 Hollow Cathodes

It has long been realized that hollow cathodes make cold cathode discharges such as neon tubes function more easily. In that case the cathode is simply a hollow tube, closed at one end, with the open end facing the plasma. It is advantageous because the electrons bouncing around inside the tube create a plasma density which is greater than elsewhere in the plasma and therefore more ions bombard the surface, which releases the secondary electrons required to sustain the discharge.

A reentrant configuration can also be advantageous for thermionic emitters. A few examples of hollow cathodes are presented in Fig. 7.19. Figure 7.19a shows the hollow cathode of Lidsky et al. (1962), simply a tungsten or tantalum tube of inside diameter 3–12.7 mm through which gas was introduced. The discharge was started as an RF discharge which caused the tube to be heated by ion bombardment. The RF power could then be turned off and the discharge sustained. Operating in an axial magnetic field of the order of 1 kgauss a 3 mm diameter thin wall tantalum tube operated stably with currents of 3–270 A with arc voltages of 30–80 volts. Interestingly, under the best operating conditions these cathodes were observed to be hottest over a length of several diameters, several diameters back of the open end. This phenomenon—maximum heating away from the end—has been observed in various types of hollow cathode. A detailed review of hollow cathodes of this type was made by Delcroix and Trindade (1974).

In a variant of the Lidsky cathode Kovarik et al. (1982) used flattened tubes to produce a broad discharge. In their case instead of initiating the discharge with RF they used an auxiliary Penning discharge to heat the cathode to emitting temperature.

The cathode shown in Fig. 7.19b is the hollow cathode of Fig. 6.2 described by Ernstene et al. (1966). The porous nickel plug, which was omitted from the more schematic Fig. 6.2, represents the mode of delivery of cesium. This porous metal served as a wick which delivered liquid cesium to the heated end. The portion of the heater surrounding the end of the wick was separately controlled so that

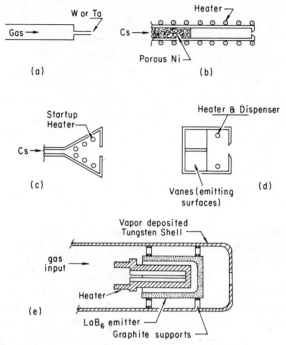

Figure 7.19. Examples of hollow cathodes: (*a*) the Lidsky cathode, (*b*) the cesiated plasma bridge cathode, (*c*) the cesiated cathode of a Cs^+ ion source plasma, (*d*) a Hull-type dispenser cathode, and (*e*) the Forrester–Goebel–Crow hollow cathode. From Goebel et al., 1981.

the cesium vapor pressure and emitting surface temperature were separately controlled. The performance of this cathode is fully described in Sec. 6.1. The emitting surface is the low work function surface created by a fractional monolayer of cesium on the inner surface of the approximately 6 mm inner diameter cylinder. The emitting properties of cesium on tungsten are described in detail in Sec. 9.4. This cathode requires temperatures of approximately 600°C so that a refractory metal cylinder is not required; stainless steel or nickel is adequate. Since cesium discharges operate at very low voltages, typically less than 10 volts, sputtering of the cathode surface is very small and, for the most part, will be the replaceable cesium monolayer layer. No lifetime limitation has been observed for this cathode. Currents of the order of 5 A have been extracted from a $1/8$ mm diameter hole and, when operated carefully, with very little cesium throughput.

The cathode of Fig. 7.19*b* was invented for the purpose of Cs^+ beam neutralization as described in Sec. 6.1. For the neutralization of other ion beams, where it was feared that cesium would be a harmful contaminant, a similar configuration has been used where the active surface was produced by coating the inside surface with a mixture of carbonates of barium, strontium, and calcium as described in Sec. 7.6 and feeding with whatever gas or vapor was appropriate to the ion beam,

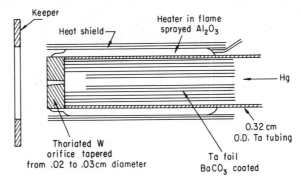

Figure 7.20. Plasma bridge neutralizer for use with mercury.

for example, mercury vapor for an Hg^+ ion source. To increase the emitting area of such a cathode Rawlin and Pawlik (1968) inserted into the cylinder a thin tantalum strip coated with the carbonate mixture and rolled into a tight spiral, as shown in Fig. 7.20. Although these small diameter (~0.5 cm) cathodes with much smaller diameter (0.2–0.5 mm) apertures were first developed as plasma bridge neutralizers, Moore (1969), Ramsey (1971), James et al. (1970), and Sovey (1981), for example, found them adequate as primary sources of electrons in ion source discharges, which will be described in Chapter 8. The particularly compact design of Moore is shown in Fig. 7.21. The heater in Moore's cathode was used for starting purposes only.

The cesiated hollow cathode shown in Fig. 7.19c is the cathode used by Sohl et al. (1966) and Speiser (1966) as the cathode of a Cs^+ ion source. All of the cesium to the ion source was fed through this cathode. The heater was for startup purposes only with ion bombardment providing an adequate heat input during steady state operation. The source operated stably with approximately 20 A at 8 volts. Because of the low voltage and the replenishable cesium adsorbed layer this cathode showed neither deterioration in performance nor any signs of erosion after thousands of hours of running.

The Hull-type dispenser cathode described in Sec. 7.7 is by its nature a hollow cathode, and an embodiment of that cathode, used by me but never previously

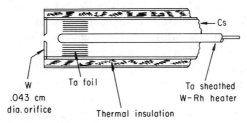

Figure 7.21. Cesium hollow cathode of Moore.

described in the literature, is shown in Fig. 7.19d. This cathode was 2.54 cm in diameter and 2.54 cm long with a 0.64 cm diameter aperture.

The hollow cathode shown in Fig. 7.19e was used by Goebel et al. (1978), Forrester et al. (1978), and Goebel and Forrester (1982) and with La_2O_3 impregnated molybdenum as the emitter by Goebel et al. (1980). The emitter in the configuration shown was 2.54 cm in diameter and 3.81 cm long and the vapor deposited tungsten shell had an inside diameter of 3.81 cm. The aperture was usually 0.64 cm. Currents of several hundred amperes could be obtained with this cathode for long periods of time.

The feature common to hollow cathodes—the reentrant structure—implies first of all that much lower power is required than for the case when the emitter surfaces face relatively cool areas. A second important advantage is that the active emitting material evaporates, for the most part, from emitting surface to emitting surface and only a small fraction of the evaporated material leaves the cathode through the electron extraction aperture. A third advantage relates to the substantially greater plasma density that is likely to exist inside the hollow cathode when gas is introduced through the cathode.

Let us examine quantitatively the advantage which results from the high density that may exist inside a hollow cathode. The ion current density from a plasma is given by Eq. (3.69) as

$$J_i = 0.344 n_0 e \sqrt{2kT/M}.$$

As discussed in Secs. 3.9 and 7.4 the limiting electron emission is likely to be limited to

$$J_e < 0.25 \, J_i \sqrt{M/m},$$

leading to

$$J_e < 0.086 \, n_0 e \sqrt{kT/m}.$$

With a plasma density of 10^{18} electrons per cubic meter, typical of intense ion sources, for reasons given in Sec. 8.1, and an electron temperature of 5 eV, we obtain

$$J_e < 1.83 \text{ A/cm}^2.$$

To take advantage of an emitting material such as LaB_6 capable of yielding current densities of 20 A/cm^2 it is necessary to have a plasma density of 10^{19} m^{-3}. This is readily obtained inside a hollow cathode leading to a requisite emitter area only one-tenth as great as it would necessarily be when exposed directly to the main plasma.

In my laboratory an interesting phenomenon was observed. In testing a cathode such as that shown in Fig. 7.19e we would customarily run the output to a flat

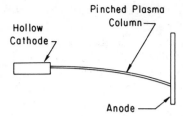

Figure 7.22. Hollow cathode test set-up.

plate serving as the anode as shown in Fig. 7.22, in order to observe the performance visually. With the large currents (up to 500 A) obtainable from this cathode the beam was pinched to a tight column which was usually observed to curve or spiral toward the anode, often whipping about due to the usual "kink" instability or "wriggling" associated with an unstabilized pinch. When used as the cathode of an ion source plasma, however, a probing of the plasma never showed significant lateral plasma density variations. The difference is easily understood. In the ion source configuration, shown schematically in Fig. 7.23, the screen electrode is usually floated so that no net current flows to this electrode. Because of one of various magnetic field configurations which serve as a barrier to direct electron flow to the anode, as discussed in Chapter 8, the electrons bounce around, gradually working their way to the anode surface. The current flow patterns must look schematically like those shown in Fig. 7.23 so that there is no possibility of pinch effects except right at the cathode orifice where it certainly is important. The pinching at the orifice is probably responsible for the fact that maximum heating occurs not at the orifice but somewhat back of the orifice, as mentioned earlier for the Lidsky cathode.

The analysis of the behavior of hollow cathodes with small openings to the main plasma poses substantial difficulties. The high density plasma inside, say, the tungsten shell of Fig. 7.19e is probably joined to the lower density ion source plasma through an even higher density plasma in the aperture. The potential in the aperture

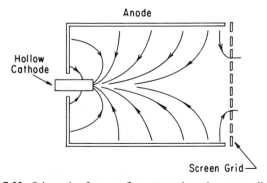

Figure 7.23. Schematic of current flow pattern in an ion source discharge.

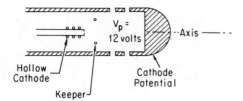

Figure 7.24. Double hollow cathode of Knauer et al.

is probably higher than the potential on either side with the three plasmas connected by two double sheaths. It would be an interesting, if somewhat difficult, problem to analyze.

In some sources, such as that of Knauer et al. (1969), whose cathode arrangement is shown in Fig. 7.24, a hollow cathode is inside another volume whose connection to the main source plasma is through a small area aperature. Knauer et al. measured a potential of only 12 volts in the enclosure surrounding the hollow cathode when the potential of the main plasma was approximately 45 volts. As a consequence the rate of sputtering was reduced relative to that which occurs when 45 eV ions can strike the hollow cathode directly.

Chapter 8

Taxonomy of Positive Ion Sources

8.1 Desired Plasma Density

Since we have considered the electrodes separately the ion source is simply a plasma generator. It has a cathode, an anode, and means of introducing gas so as to produce a plasma of some desired density. Magnetic fields are frequently employed in a variety of ways as well. Before considering the various configurations of these elements it is important to note the desired range of plasma density.

The ion current density available from a plasma is given by Eq. (3.69) as

$$J = 0.344 n_0 e \sqrt{2kT/M}. \tag{3.69}$$

For ion source plasmas, for quiescent laboratory plasmas generally, the electron temperature T is usually about 5 eV. The ion mass is usually determined by the purpose of the ion source. It is the quantity n_0, the plasma density, which primarily controls the available current density.

The current density is also limited by the geometry of the plasma and accel electrodes (see Fig. 1.1). As an approximation consider these to be highly transparent with apertures smaller than the spacing d between these electrodes. With initial energies very small compared to the energy to which the ions get accelerated the current density must then be given by the Child equation (2.8),

$$J = (4\varepsilon_0/9) \sqrt{2e/M} \, V^{3/2}/d^2, \tag{2.8}$$

where V is the voltage across the accel gap taken to be of width d. Equate the right hand side of Eqs. (2.8) and (3.69), set $\varepsilon_0 = 8.85 \times 10^{-12}$ F/m and $e = 1.6 \times 10^{-19}$ C to obtain

$$n_0 = 7.15 \times 10^7 V^{3/2}/(d^2 \sqrt{kT/e}), \tag{8.1}$$

where all quantities are in SI units. With kT/e taken as 5 volts Eq. 8.1 becomes

$$n_0 = 3.2 \times 10^7 V^{3/2}/d^2. \tag{8.2}$$

The desirability for large n_0 then translates into a need for small d. As a practical matter d is limited, at voltages less than approximately 40,000 volts, by structural and thermal stability problems. A convenient if somewhat arbitrary choice is $d = 4 \times 10^{-3}$ m. This leads to a value of

$$n_0 = 2 \times 10^{12} V^{3/2}, \tag{8.3}$$

where n_0 is in inverse cubic meters and V is in volts. For example, for $V = 5000$ volts this crude treatment leads to

$$n_0 = 7 \times 10^{17} \text{ m}^{-3}$$

and at 40,000 volts it leads to

$$n_0 = 1.6 \times 10^{19} \text{ m}^{-3}.$$

At voltages greater than 40,000 volts the limitation on d is more on the field in the accel gap than on structural considerations. The field strength that can be used before breakdown occurs with unacceptable frequency depends on many conditions and is observed to decrease as voltage increases. For the purpose of the rough exercise undertaken here take the gap field as $V/d = 10^7$ V/m for voltages >40 kV. Equation (8.2) then yields

$$n_0 = 3.2 \times 10^{21}/\sqrt{V}. \tag{8.4}$$

This yields the same value $n_0 = 1.6 \times 10^{19}$ m^{-3} at 40,000 volts that I found for the low voltage case. In general we can say from this rough treatment that we desire ion source plasmas to have densities in the range

$$5 \times 10^{11} \text{ cm}^{-3} < n_0 < 2 \times 10^{13} \text{ cm}^{-3}. \tag{8.5}$$

The lower end of this range, 5×10^{11} to 10^{12} or even 2×10^{12} cm^{-3} is accessible without great difficulty. However, the upper end poses substantial problems. To illustrate the difficulties consider a hydrogen ion plasma of density of 10^{19} m^{-3} and an electron temperature of 5 eV. Equation (3.69) yields a current density of

1.7 A/cm^2. As shown in Fig. 3.11 the plasma would be expected to ride about 18 volts above any floating electrode, which would usually include the plasma electrode. If the plasma contains a small percentage of primary electrons the floating potential would be greater, but even this figure leads to a power density of 31 W/cm^2. That much power requires that a coolant be circulated through the electrodes, a step that substantially complicates the construction of high transparency extraction electrodes. Because the ion current density decreases as ion mass increases the power loading is much less of a problem for massive ions such as Hg$^+$, Cs$^+$, or Xe$^+$.

8.2   Single Aperture Sources, No Magnetic Field

This book is oriented toward ion beams of very high poissance, the normalized perveance quantity defined by Eq. (2.116) or (2.122). Such beams necessarily require multiaperture extraction systems, including the fine mesh system of Sec. 5.7. Because such large extraction area sources have evolved from single aperture sources it is helpful to start with some description of single beam sources.

It is not my intention to make this a comprehensive catalog, but rather to present a few types which lead toward large multiaperture sources. Some of the sources discussed here do not quite fit this description but illustrate the generality of configurations which can serve as sources of ions. Magnetic fields play a key role in ion source performance, and this feature is used as a guide to categorization.

A.   Canal Ray Discharge Sources

When a voltage is applied between two cold electrodes a steady state glow discharge can be produced. This discharge is characterized by a moderately field-free plasma region at a slightly positive potential relative to the anode and a large voltage drop across a cathode sheath. The electrons which sustain the discharge are the secondary electrons released from the cathode by ion bombardment.

The voltage required to sustain this discharge can be anything from 400 to 20,000 volts depending on the gas pressure, necessarily greater than about 10^{-3} torr, the nature of the gas, the electrode materials, and the dimensions of the apparatus. The dimensions must be large if pressures are to be kept small. A hole in the cathode long compared to its diameter would then yield a stream of ions to which the term canal ray (from the German *Kanalstrahlen*) was applied. Typically the hole might be one or several millimeters in diameter and several centimeters long. Although the energy spread from a source of this type may be large, the spread in angles can be small. Thomson (1910, 1911) was able to use ions from a canal ray source for mass analysis with no further acceleration. Figure 8.1 shows the version of a canal ray source used by Oliphant and Rutherford (1933). This source ran with a discharge voltage of 20 kV and the emerging ions then accelerated through an additional 200 kV.

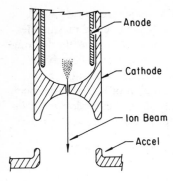

Figure 8.1. Canal ray ion source of Oliphant and Rutherford.

B.   Capillaritron

I jump from the oldest type of source to the most recent source, which, surprisingly, is the simplest configuration of any source. The capillaritron, described by Mahoney et al. (1981), is shown in Fig. 8.2. In this source a vapor or gas of any substance is passed through a nozzle with a very small diameter orifice, 50 μm or

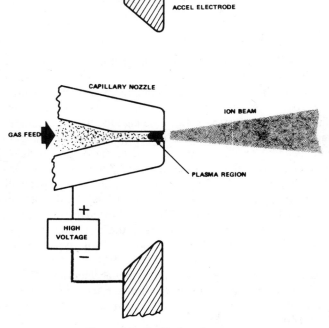

Figure 8.2. Capillaritron ion source.

less. This nozzle is in the center of a hole in the grounded accel electrode. For an appropriate feed rate the application of a voltage greater than about 400 volts results in a virtually noise-free ion beam. Tests with argon, helium, hydrogen, oxygen, and xenon have been reported and the source appears capable of operating with any gas. Current–voltage characteristics for argon are shown in Fig. 8.3. The current emerges into a narrow cone and Mahoney et al. report an argon brightness greater than 6 mA/steradian for the source.

The analysis of this extraordinary source is based on a self-consistent picture of a plasma inside the fine capillary. According to this picture the plasma is maintained by a small current of high energy electrons from the accelerating field region back into the capillary but with the bulk of the ionizing electrons coming from the walls of the capillary. These are released in copious amounts by the large number of excited atoms which reach the walls. Because the inside diameter of the tube is so small a large fraction of all excited atoms will strike the walls before they can radiate. The transfer of excitation into electron energy by this type of collision is an efficient process and Mahoney et al. show that the ion current density, the atom density, and the tube radius are reasonably consistent with the Tonks–Langmuir theory of a plasma presented in Secs. 3.5, 3.7, and 3.10.

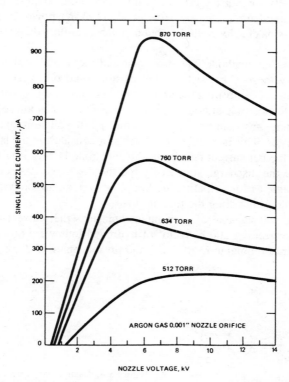

Figure 8.3. Characteristics of a capillaritron for various feed pressures.

Two serious deficiencies in the operating parameters of the capillaritron limit is application. The first of these is the large gas flow. Perel et al. (1981) show, for example, an argon ion current of 0.6 mA at 10 kV with a gas flow rate of 2.04 SCCM (standard cubic centimeters per minute). Since 1 SCCM can readily be shown to be equivalent to a current of 71.7 mA, assigning one electronic charge to each atom, this converts to a gas utilization efficiency of 4×10^{-3}. The second serious defect is a consequence of this. The energy spread in the beam, presumably as a result of charge transfer during acceleration, is essentially flat, covering the entire range from zero to the maximum. This feature of the capillaritron was measured subsequent to the published papers on the source.

C. Low Voltage Capillary Arcs

The simplest, lowest power means of creating dense plasmas is through the use of a thermionically emitting cathode and an anode immersed in the gas to be ionized. With ample emission a discharge readily forms at a voltage of, usually, 25–80 volts depending on the gas and the configuration. The voltage appears primarily across the cathode sheath and is usually in a range where the ionization cross section is a maximum. A steady state plasma from which ions can be extracted readily forms. By forcing the current to flow through a small diameter capillary, as in the capillary arc source of Tuve et al. (1935), shown in Fig. 8.4, large increases in plasma density could be obtained. The capillary diameter in this source was 3.5 mm.

In this source the cathode was grounded and the anode connected to a +110 volt DC supply through a resistor. The resistor could be varied from 40 to 500 ohms as a means of controlling the arc current. The arc body, which I have labeled the intermediate electrode although this term was not used in ion source discharges until much later, was connected to the same +110 volt terminal but through a 1000 ohm resistor. This is a high enough resistor considering the large area of this electrode that the net current density to this electrode is very small. This connection will make the discharge start easily but the discharge characteristics will be virtually the same as for a floating arc body, or even an arc body of an insulating material capable of handling the thermal load.

This source was capable of such intense plasmas that current densities to the aperture area exceeding 200 mA/cm^2 of hydrogen ions were observed. If we assume an electron temperature of 5 eV and an ion mass of 2 we see, using Eq.

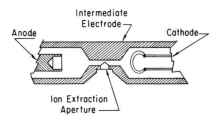

Figure 8.4. Capillary arc source of Tuve, Dahl, and Hafstad.

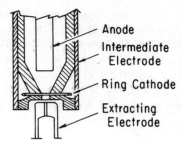

Anode

Intermediate
 Electrode

Ring Cathode

Extracting
 Electrode

Figure 8.5. Zinn ion source configuration.

(3.69), that they had a plasma density of the order of 1.5×10^{12} cm^{-3}. They actually drew currents as large as 1.5 mA from a probe canal, a 1 mm diameter hole in an accel electrode of length much greater than 1 mm.

An interesting variant of the capillary arc source is the configuration used by Zinn (1937) and shown in Fig. 8.5. In this variation of the capillary arc the electrons from a ring cathode (a tungsten filament was found by Zinn to be most effective) converge through an annular aperture 1 mm thick to a concentric hole 3 mm in diameter and then bend through 90° to a water-cooled anode. The extraction, as seen, is directly opposite the anode. With this source Zinn was able to extract as much as 4.3 mA from a hole in the extraction electrode 1 mm in diameter and 6 mm long.

The Unoplasmatron ion source of Ardenne (1956), shown in Fig. 8.6, is very similar to what the Zinn source would be if the cathode and anode positions were simply reversed. This is not precisely so, insofar as the ions are drawn out through the anode and not through the intermediate electrode as in the Zinn source. In Ardenne's Unoplasmatron the hole in the intermediate electrode had a 5 mm diameter with the extraction hole in the anode of diameter 1–2.5 mm. This source was also built by Ardenne with long slit extraction. In the latter case the slits in the intermediate electrode and the anode were each 0.3 cm × 14 cm.

In Ardenne's picture of the manner in which this source functions he envisions a low density plasma surrounding the cathode and a high density plasma near the

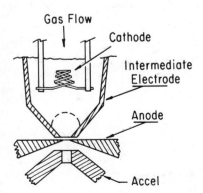

Gas Flow

Cathode

Intermediate
Electrode

Anode

Accel

Figure 8.6. Unoplasmatron ion source of Ardenne.

extraction orifice at a potential close to the anode. These two plasmas are joined by a double sheath along the surface shown by the dashed line in Fig. 8.6. The high density plasma near the aperture is maintained by the electrons accelerated across the double sheath. This surmise as to the nature of the phenomenon was sustained by a source constructed to provide a view into the region around the extraction aperture and by probe measurements of the plasma. With a discharge current of 1 A at 70 volts Ardenne obtained a beam current of 7.8 mA at 60 kV. He measured a 19% mass utilization efficiency.

8.3   Single Aperture Sources, Magnetic Fields

A.   Oscillating Electron, Cold Cathode (Penning) Sources

In the geometry illustrated in Fig. 8.7 the cathodes are two parallel disks normal to a magnetic field of several hundred gauss. The anode lies in a plane halfway between the cathodes with a hole in the center. Electrons released from a cathode, constrained to move along magnetic field lines, get accelerated toward the median plane, pass through the hole in the anode, and then get reflected from the opposite cathode. These oscillating electrons must eventually work their way to the anode by virtue of collisions and microinstabilities in the plasma which forms when sufficient ionization occurs. With a cold cathode the electrons are released by ion bombardment and a steady state glow discharge will form at pressures much lower than possible with comparable dimensions in the absence of a magnetic field. With a voltage of approximately 2 kV it was found that a discharge could be maintained at pressures as low as 10^{-5} torr. The first application of this discharge by Penning (1937) was as a vacuum gage with advantages relative to the conventional ionization gage in its simplicity and absence of a thermionic cathode. In this application the ion extraction hole shown in Fig. 8.7 is not present.

With a hole in one of the cathodes ions can be extracted and accelerated to any desired energy with substantial advantages over previous sources in the combination of small size and low pressure required, as first demonstrated by Penning and Moubis (1937).

The source voltage depends critically of the nature of the cathode material,

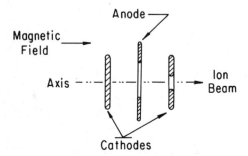

Figure 8.7. Penning ion source.

being lowered by the use of materials with high secondary emission yields. Backus (1949) studied the variation with cathode material and found that the characteristic operating voltages varied from 3600 volts for nickel cathodes to 350 volts for aluminum and 280 volts for beryllium, as long as the normal oxide coating on the aluminum and beryllium was not disturbed. By using a hollow aluminum cathode Glazov et al. (1964) ran a Penning ion source in a pulsed mode with a discharge voltage of only 150 volts.

B. Oscillating Electron, Hot Cathode (Finkelstein) Sources

The use of a thermionic cathode in the oscillating electron configuration made it possible to extract currents limited only by considerations of cathode lifetime and ability of electrodes to handle the current, and with increased levels of gas utilization efficiency. The first such source, that of Finkelstein (1940), is shown in Fig. 8.8. With 0.63 cm diameter oscillating electron and extraction apertures, and a magnetic field of 360 gauss, Finkelstein reported currents as high as 150 mA of hydrogen ions. With the 3 kV available to him, the maximum current which could be accelerated into an ion beam, as given by Eq. (5.2) with $\chi = 3.85 \times 10^{-8}$ $A/V^{3/2}$ for ions of atomic mass 2, is only 6 mA. For currents larger than this Finkelstein observed wide angle spreading during ion acceleration so that only a small fraction of this current would continue through successive holes in his accel electrodes.

The sources of Bailey et al. (1949) shown in Fig. 8.9, of Abele and Meckbach (1959) shown in Fig. 8.10, and many others, are examples of Finkelstein ion sources, that is, hot cathode sources in which electrons oscillate along magnetic field lines in an approximately uniform field, with ions drawn out through an aperture opposite the cathode.

The use of an anticathode, as in Fig. 8.10, fixes the boundary of the surface at which electrons are turned back and from which ions are accelerated, but the source of Bailey et al. omits such an electrode. With an arc voltage almost certainly under 100 volts (not specified by Bailey et al.) and an accelerating voltage of 200 kV the cathode potential surface can readily be pushed back into the hole in the anode

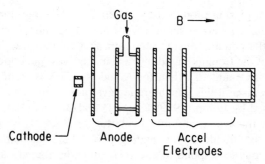

Figure 8.8. Hot cathode oscillating electron ion source of Finkelstein.

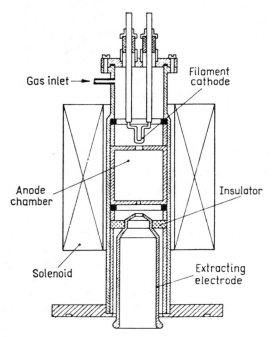

Figure 8.9. Oscillating electron source of Bailey, Durkey, and Oppenheimer. From Valyi, 1977.

chamber to produce the well focused beam they observed. With or without the anticathode it is an oscillating electron source.

Finkelstein describes the electrons as compelled by the negative potentials at each end of the plasma column to move out laterally. This creates, according to him, a saddle point potential distribution. Specifically, he imagines that the potential variation transverse to the magnetic field must show a minimum along the axis of the plasma in order to force the electrons out laterally. This would then prevent ions from moving out laterally and all ions would either flow to the cathode or to the extraction aperture depending on which side of the maximum the ions were produced.

This view of a plasma produced by electrons oscillating along magnetic field lines appears to have received support from measurements of Kistemaker (1955), who found that the potential inside such a plasma column was, in fact, negative. The magnitude of the troughs found by Kistemaker was much greater for the electronegative gases, O_2 and Cl_2, than for A and N_2. This is readily understandable in terms of microinstabilities which are always present in a plasma in a magnetic field. As discussed by Bohm (1949, Chap. 1) and Backus (1949), electrons have a relatively high transverse mobility by virtue of $\mathbf{E} \times \mathbf{B}/B^2$ drifts, where the electric field E has a randomly fluctuating part. These fluctuations are too rapid for the negative ions to respond to and these must be forced out laterally by the

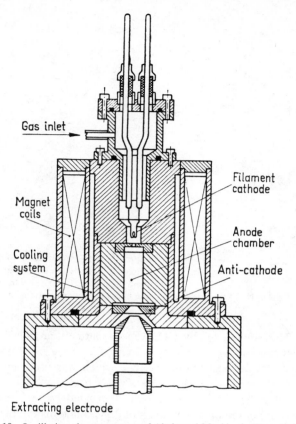

Figure 8.10. Oscillating electron source of Abele and Meckbach. From Valyi, 1977.

radial field. The existence of negative ions in a Finkelstein source is likely to enhance the source performance by improving the funneling of positive ions to the extraction aperture.

C. Transverse Extraction (Calutron) Sources

The trough produced by electrons oscillating along magnetic field lines might be expected to make transverse ion extraction from oscillating electron sources inefficient. In fact transverse extraction sources have been highly successful.

An example of a source with transverse extraction is the Calutron* source seen

---

*Calutron is named for the University of California where this source was developed for the wartime (1942–1945) project for the separation of U^{235} from the dominant U^{238} isotope. I am not aware of any description of this source in the archival literature. The only two books on the subject that I am cognizant of, Guthrie and Wakerling (1949) and Wakerling and Guthrie (1949), avoid detailed descriptions of the source as if it still had been regarded as classified information in 1949.

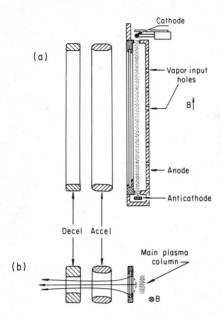

Figure 8.11. Representation of the Calutron ion source by (*a*) a section parallel to the magnetic field along the direction of beam extraction and (*b*) a section normal to the magnetic field.

in Fig. 8.11. This source, fed by UCl$_4$ vapor, routinely produced about 100 mA of ions from an exit slit approximately 1.3 cm $\times$ 25 cm. This is about as much as one could expect at the accelerating voltage which was used, approximately 35 kV, based on Eqs. (5.3) and (2.116), for a poissance of about 0.4, very high for the small (± 0.1 radian) divergence angle demanded by the isotope separator. (These numbers are based on an ion mass of 238 atomic mass units. Although the actual beam was primarily U$^+$ it also contained large numbers of Cl$^+$, UC$_x^+$, and many doubly charged ions. The calculation is therefore a rough one but does indicate that the source was able to supply, transverse to the field, an ample supply of ions.) A virtually identical source was developed by Kistemaker et al. (1955).

The anticathode, shown in Fig. 8.11, an unheated tungsten plate at cathode potential, improved performance somewhat but was not an essential element. The source worked approximately as well when the plasma column was terminated, on the end opposite the cathode, by anode potential.

In these sources the slit defining the primary electron stream is narrow so that electrons readily drift to regions where they can move along field lines to electrodes at anode potential. For this configuration Bohm et al. (1949) showed that the plasma potential is positive relative to the anode so that the efficient transverse collection of ions is readily understood.

It might be imagined that the flow along the field lines would be so rapid compared to plasma flow across the field that the analysis of Sec. 3.5A would apply, leading to a variation in the plasma density along the length of the column as shown in Fig. 3.5. In fact, at pressures for which the mean free path for primary electrons is of the order of the length of the arc or greater the plasma shows no such density variation. The current density and the beam spreading appear to be uniform along the length of the arc. Although the plasma flow along field lines is certainly more rapid than across field lines, the small size of the lateral dimensions compared to the length dictate that the primary ion drift is transverse to the field, except close to the end of the column.

D.   Duoplasmatron Ion Source

In the duoplasmatron Ardenne (1956) caused the plasma from a unoplasmatron to be brought to an intense sharp focus at the extraction aperture in the anode by a strong magnetic lens (Fig. 8.12). He achieved this by making the intermediate electrode and anode of iron and producing a longitudinal field with the coil shown in Fig. 8.12. The intermediate electrode was made sufficiently thick so that the region surrounding the cathode was essentially free of magnetic fields, making possible the unoplasmatron action in the region between the cathode and intermediate electrode. As the electrons from the cathode area move toward the anode they encounter the convergent magnetic field, shown qualitatively in Fig. 8.13, causing an extremely intense small diameter plasma at the extraction aperture in the anode. With this source Ardenne achieved a steady state beam current of 80 mA of H^+ ions from a 1.2 mm diameter aperture with an extraction voltage of 60 kV. His mass utilization for the 120 volt, 2 A discharge which produced the 80 mA beam was a remarkable 95%.

In a duoplasmatron developed by Kelley et al. (1961) the ferromagnetic anode proved unnecessary. Their geometry was similar to that of Ardenne but had an

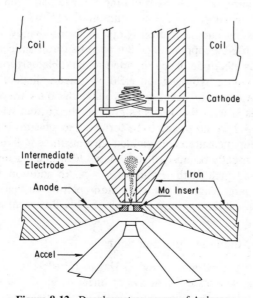

Figure 8.12. Duoplasmatron source of Ardenne.

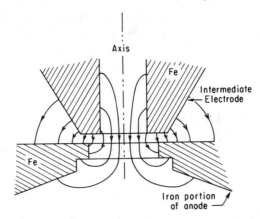

Figure 8.13. Qualitative drawing of magnetic field lines between the pole pieces (intermediate electrode and anode) of a duoplasmatron.

anode of water-cooled copper with a tungsten insert, instead of the iron anode he used. The increase in magnetic field from a low value in the cathode region to an intense value at the anode aperture, which is the essence of the duoplasmatron, does not require an iron anode and Kelley et al. measured a field of 4 kgauss at the position of their anode aperture. They achieved hydrogen ion (70% H_2^+) beam currents of 300 mA at 80 kV. Since the poissance (see Sec. 5.2) represented by these figures is 0.34, they are at the space charge limit of a single aperture accelerating system and it is possible the source might have been able to supply substantially greater currents if the accelerating voltage had been higher. Their gas efficiency was estimated to be 92% at a current of 215 mA.

Another interesting duoplasmatron is due to Demirkhanov et al. (1964). Like Kelley et al. they used a copper anode but added an iron anticathode immediately downstream of this electrode. They reported remarkable performance figures. Unlike Kelley et al. whose hydrogen ion yield was predominantly H_2^+, they claimed an 85% proton output. At an accelerating voltage of 30 kV they reported a current of 1.5 A, which is at least five times as great as the current which considerations discussed in Sec. 5.2 would permit to be formed into a beam. However, they point out that Faraday cup measurements yielded currents three to five times greater than obtained calorimetrically because of secondary electron emission. It was not made clear which method yielded the figure of 1.5 A. In addition, their Faraday cup subtended a very large angle at the extraction aperture so that the source plasma might have penetrated the aperture to yield a highly divergent ion beam.

E. Single-Ring Magnetic Cusp Ion Source

Although the source of Brainard and O'Hagan (1983; Fig. 8.14) evolved from large multiaperture sources using arrays of linear magnetic cusps, as discussed in Secs. 8.9–8.12, this interesting source falls in the category of single aperture

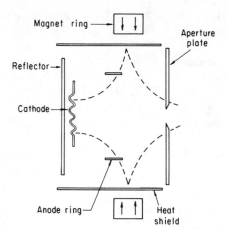

Figure 8.14. Single ring magnetic cusp ion source of Brainard and O'Hagan. The reflector, heat shield, and aperture plate were electrically floated.

sources. In this source a ring of strong permanent magnets provided a field of 0.2 tesla at the circular cusp inside the heat shield and point cusps at the cathode and aperture plate. A null exists in the magnetic field along the axis.

All electrodes were floated except the cathode and anode. The reflector electrode ran close to cathode potential, the aperture plate close to anode potential, and the heat shield approximately halfway between. The plasma potential, as is normal, assumed a potential close to anode potential and the floating aperture plate, which would normally ride up near cathode potential in order to repel primary electrons, did not need to serve that function. The converging magnetic field effectively reflected primary electrons so that the region near the aperture plate contained mainly thermalized electrons, accounting for its floating close to anode potential.

The formation of a plasma which had a large density of primary electrons in the half closest to the cathode and a low density of energetic electrons in the half closest to the aperture plate had a helpful effect as far as the makeup of the deuterium ion beam with which Brainard and O'Hagan were working. As shown by Ehlers and Leung (1981), a high density of energetic electrons is responsible for efficient production of D_2^+ in the rear of the source, and the passage of the D_2^+ ions through the thermalized plasma is responsible for the decomposition to atomic ions.

A particular goal of Brainard and O'Hagan was the development of a source that would operate at low pressure, and their source required only 0.25 Pa (1 Pa = 7.5 × 10^{-3} torr).

8.4   Broad Multiaperture Sources, No Magnetic Field

It is feasible to approach the problem of creating a plasma of high density and large cross section with a simple box, as shown in Fig. 1.1, in which one or several emitters are held at a negative potential relative to the walls. Aside from power loading of the walls at high plasma densities there is an inherent, but not insurmountable difficulty, with this approach.

The difficulty relates to the expected decrease in plasma density toward the boundaries of a plasma, as developed in Sec. 3.5 and illustrated in Fig. 3.5. This difficulty was overcome in the source of Ehlers et al. (1973), shown in Fig. 8.15a. The discharge occurred in a cylinder whose length was less than half the diameter. The cathode consisted of 20 tungsten hairpins with a total emitting area of 34 cm² radially arranged near the cylindrical surface. The cylindrical surface alone served as the anode with both the front surface (extraction area) and rear surface elec-

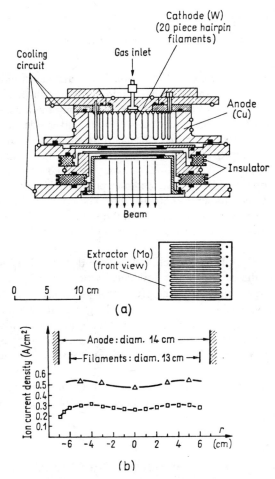

Figure 8.15. Drawing of (a) ion source of Ehlers et al. (1973) and (b) current density profiles for deuterium at two different discharge currents. Figure from Valyi (1977), with correction for mislabeling of current density profiles.

trodes floating. The peripheral positioning of the cathode and anode resulted in an ion current density profile which was measured by movable probes to be flat to $\pm 6\%$ over a diameter of 12 cm and, as seen in Fig. 8.15b, actually showed a minimum at the center.

Gas utilization of so short a source is necessarily poorer than for a longer source of comparable plasma density and electron distribution, although Ehlers et al. report a 25% gas utilization at a deuterium ion current density of 0.5 A/cm^2. Their source was used for the production of energetic neutral beams, and neutral gas downstream of the electrodes was used as the medium for charge transfer of the energetic ions to neutrals. As discussed in Sec. 5.6, the effect of neutral gas in the accelerating gap results in serious drain currents and parasitic beam components. Even where a large neutral density is required downstream for charge transfer it would be advantageous, as pointed out by Forrester and Dawson (1978), to make the gas density in the region of the accelerating electrodes very small and introduce the gas as a localized jet downstream of the ion source.

Another disadvantage of this source is the large power loading both from the filaments and from the anode loading. The source was therefore used in a pulsed mode with 1 second required for the filaments to reach their operating temperature of 3200 K, and 0.03 second pulses on the discharge and electrode voltage. For long pulses or continuous operation this source would require, at 0.5 A/cm^2 ion current, the circulation of coolant through the plasma grid as well as the anode.

8.5   Broad Oscillating Electron (Kaufman) Sources

The application of trapped oscillating electrons from a thermionic emitter with ions extracted in the direction of magnetic field lines, as in the Finkelstein ion source, was first applied to broad sources with multiaperture extraction by Kaufman and Reader (1960) to the creation of a mercury ion source for ion propulsion applications. This type of source, illustrated in Fig. 8.16, has been widely applied, for example, by Sohl et al. (1966) to create Cs$^+$ ions and by Abdelaziz and Ghander (1967) as a source of H$^+$ ions.

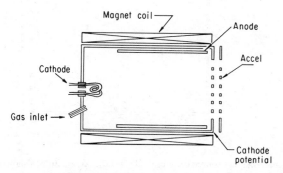

Figure 8.16. Kaufman-type ion source.

There are numerous variants of the Kaufman source. The electron emitting cathode can take on any of the forms discussed in Chapter 7 and the magnet coil can be replaced by an arrangement of permanent magnets and soft iron. The optimum magnetic field is inversely proportional to the anode diameter and Sohl et al. (1966) find the product of anode radius and optimum field strength to be 55 gauss-cm for their cesium discharge. The optimum field strength is generally several times as great for other gases which require higher arc voltages but in any case the required field of no more than a few tens of gauss is readily obtained with permanent magnets and soft iron as shown in Fig. 8.17.

Not only was the replacement of the electromagnet by a permanent magnet a simplification, it also resulted in an improved performance as measured by a decrease in the energy per ion and an increase in the mass utilization efficiency. This was attributed to the fact that the magnetic field lines, for the source shown in Fig. 8.17, diverged toward the screen electrode more than for the electromagnet version of this source. Optimization studies by Bechtel (1968) led to the configuration in Fig. 8.18 in which the magnetic field very strongly diverged toward the extraction area. The improvement in performance associated with the divergence of the magnetic field is usually attributed (e.g., Brewer, 1970) to the achievement of an arrangement by which primary electrons from a cathode on axis can reach all parts of the ion extraction area. Another point made by Sohl et al. is based on the average force,

$$F_z = -\mu \partial B / \partial z, \tag{8.6}$$

which a converging or diverging field exerts on a charged particle gyrating about a magnetic field line. In this equation, developed, for example, in Chen (1974), μ is the magnitude of the magnetic moment, which is an invariant of the electron

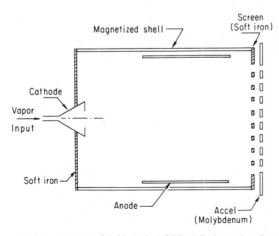

Figure 8.17. Permanent magnet source of Sohl et al. (1966). All electrodes shown except the anode and accel are run at cathode potential.

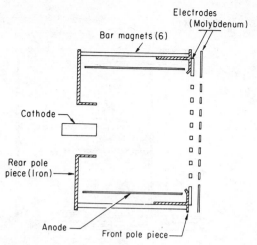

Figure 8.18. Configuration of a permanent magnet divergent field Kaufman source as optimized by Bechtel (1968).

motion. This force on the electrons toward the screen will establish a potential gradient that can be expected to accelerate ions toward the screen. The consequence should be a more efficient collection into the beam of the ions formed within the volume of the plasma than would be expected for a uniform field.

A difficulty with Kaufman-type sources relates to the decrease in density from the center of the plasma to the outer radius. This variation, usually by a factor of about 3, requires some perveance matching. As mentioned in Sec. 5.5, this difficulty has been attacked by a gradation in the extraction aperture and by a radial variation in the screen to accel spacing. These solutions have proven adequate for ion propulsion where very small beam spreading is not essential. For applications such as ion sources for the generation, through charge transfer, of neutral beams for injection into magnetically contained fusion plasmas, angular divergences of the order of $\pm 1°$ are required and adequate perveance matching would prove extremely difficult. For such sources a plasma that is highly uniform over the extraction area is desired.

8.6 Radial Field Source

The uniformity of ion emission along the length of the Calutron source described in Sec. 8.3C suggests that uniformity over a large extraction area might be achieved with a radial magnetic field with ions extracted normal to the field lines. However, if the magnetic field were strictly radial, the field strength would fall off as $1/r$ and a varying plasma density would be expected. This could, in principle, be

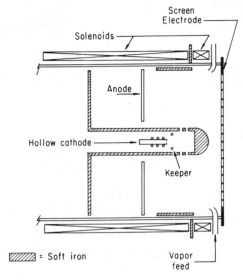

Figure 8.19. Radial magnetic field ion source of Knauer et al.

compensated for by having the lines compress in the z direction as the radius increased.

Knauer et al. (1969) set out to construct a radial field source with the geometry shown in Fig. 8.19. A major departure in configuration, besides the arrangement of the iron pole pieces, is the transformation of the usually cylindrical anode into a flat plate parallel to the screen electrode. They reported some outstanding results. The energy consumption of 190 eV per mercury ion at 90% propellant efficiency represented an improvement over previous Hg^+ ion sources and the goal of improved uniformity was achieved. In fact, under certain conditions the ion current density increased slightly toward the edge of the beam as measured 1 cm downstream of the accel electrode.

Figure 8.20 is the magnetic field pattern which Knauer et al. observed with iron filing tracings. It is clear that their source is not a radial field source in the sense I discussed in the opening paragraph of this section. Rather, it is a Kaufman source with highly divergent field lines. I judge that there are two features which are the primary causes of the excellent performance of this source. The first is the highly divergent nature of the field which tends, as discussed in the previous section, to deliver a large fraction of the ions generated to the extraction area. The second is the annular nature of the apertures in the pole piece surrounding the cathode* and the fact that they connect, along field lines, to the outer portion of the extraction area. Since primary electrons would have some difficulty reaching portions of the plasma close to the axis, it is not surprising that the plasma density is not peaked

*This cathode is discussed in the last paragraph of Chapter 7.

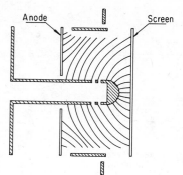

Figure 8.20. Magnetic field pattern for the source of Knauer et al.

in the center and is even found, under certain conditions, to be peaked toward the outer radius.

It is not likely that the planar form of the anode has anything to do with the performance. It may be presumed that a cylindrical anode *which connects to the same magnetic field lines* would not change the performance of this ion source.

8.7 DuoPIGatron

The extraordinary success of the duoplasmatrons of Ardenne (1956), Kelley et al. (1961), and Demirkhanov et al. (1964) led Davis et al. (1972) to investigate whether they could use that source in such a way that the current limitation imposed by a single circular extraction aperture could be overcome. Accordingly, they expanded the plasma from a duoplasmatron into a larger volume containing a diverging magnetic field terminating in multiaperture accel–decel extraction electrodes. In the larger volume the oscillating electron feature of the Penning (1937) Ion Gauge (PIG) was operable, and this new source was given the name Duo-PIGatron to suggest the combination of the duoplasmatron and the PIG configurations.

After initial successful results Davis et al. (1975) built their DuoPIGatron II ion source (Fig. 8.21). Both the intermediate electrode and the electrode labeled the anticathode are connected to the anode through high resistances, 1000 and 200 ohms, respectively. During operation this forces them to their floating potential, which is very close to the cathode potential.

A surprising identity appears not to have been recognized. A comparison of Figs. 8.18 and 8.21 discloses the duoPIGatron to be, in fact, a divergent field Kaufman source with a hollow cathode. The emitting tungsten coil plus the surrounding intermediate electrode form a hollow cathode. This cathode, like the cathode in Fig. 8.17, operates in a region of very weak field. In both cases the electrons from the cathode enter a region of relatively strong magnetic field which diverges toward the extraction area. That the virtual identity of the two sources

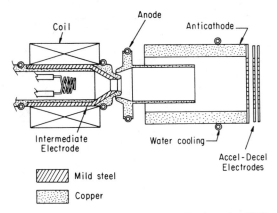

Figure 8.21. DuoPIGatron ion source (Davis et al., 1975).

has not been recognized is due in part to the separation of the ion propulsion community responsible for the Kaufman source and its variations from the fusion plasma neutral beam community responsible for the duoPIGatron, and in part because the sources have such different appearances. The difference in appearance, in large measure, is due to the much greater power dissipation encountered in a source of light ions such as H^+ or D^+ compared to a source designed for heavy ions such as Hg^+ or Cs^+, which are essential for propulsion.

To expand on the latter point I note that each type of source is pushed to the point where the current extracted is limited by the currents that can be handled by the accelerating electrodes. Although other factors enter, such as the desired ion energy, the ion mass is a dominant consideration and the available current density is proportional to $1/\sqrt{M}$. Approximately the same current density bombards all plasma boundaries. The potential of the plasma relative to boundaries would normally be at least that given by Fig. 3.11. This suggests that the high atomic mass plasmas would ride at a higher potential relative to the boundary than the light ion plasmas. However, because ionization potentials for light atoms (e.g., 13.6 volts for H, 24.6 volts for He) are large compared to those for heavy atoms (e.g., 3.9 for Cs, 10.4 for Hg), the discharge voltage and temperatures in light gas discharges are sufficiently greater than for heavy atom discharges that the plasma will generally ride at a higher potential relative to its boundaries.

For electrodes which are approximately at cathode potential the discharge voltage appears between the electrode and the plasma. The greater current density and greater potential difference between the plasma and its boundaries which exists for light ion plasmas results in the necessity of more massive construction and more effective cooling than required for heavy ion plasmas and make sources which are fundamentally similar appear quite different.

Like the Kaufman sources the duoPIGatron tends to yield a current density which is peaked at the center. This objection can be largely overcome by placing

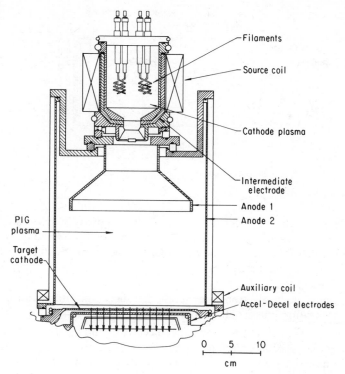

Figure 8.22. DuoPIGatron ion source (Sterling et al., 1977).

a tungsten button on the axis of the source in the transition region between the intermediate electrode and the anode, thereby producing an annular opening between the cathode plasma and the main plasma. It is exactly analogous to the uniformity produced by the annular aperture in the cathode of the radial field source described in Sec. 8.3C.

The tungsten button is seen in a later version of the duoPIGatron shown in Fig. 8.22. In this version the electrode which was labeled anticathode in Fig. 8.21 has been transformed to anode 2. Both anode 1 and 2 are connected to the positive end of the discharge supply through low resistances, tenths of ohms, so that they ride at positive potential. An auxiliary coil near the ion extraction area is adjusted to provide a balance of current between the two anodes, which yields optimum performance.

8.8   An Idealized Source

I take the liberty of giving my imagination free rein for the purpose of concocting the perfect ion source, and the source I come up with is shown in Fig. 8.23. The

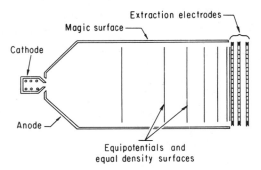

Figure 8.23. Idealized ion source.

magic surface, which comprises the anode of my imaginary source, is required to have the following fanciful properties. It should be totally reflecting of ions and fast electrons, that is, accepting only thermalized electrons. Such a source would exhibit several desirable properties.

1. Since only slow electrons can reach the anode the utilization of primary electrons for creating ions would be optimized.

2. Since all ions, except for those which necessarily flow to the cathode in order to satisfy cathode sheath requirements, are forced to flow to the extraction area, the power required for a given ion current would be minimized.

3. Except near the cathode, the ion flow would be one dimensional. Associated with this flow the equipotentials would be planes perpendicular to the axis. Along these equipotentials the plasma density would necessarily be constant and the current density would be uniform over the ion extraction area. The potential variation from the screen electrode toward the cathode would vary as shown in one of the curves of Fig. 3.9, with the grid at the $x/a = 1$ position. Neutral atom density would be expected to be inversely proportional to source length, as discussed in Sec. 3.10.

8.9   Magnetic Boundaries

Approximations to the magic surface of Fig. 8.23 are possible and, using arrays of permanent magnets, have proven feasible and effective. These boundaries have their antecedents in the so-called picket fence plasma boundary illustrated in Fig. 8.24 and discussed by Tuck (1954). In general, the picket fence is an array of parallel conductors in which the current varies from wire to wire in such a way as to produce a linear cusped field facing the plasma. The current from wire to wire may vary sinusoidally, there can be several with current in one direction followed by the same number with the same current in the opposite direction, or, as shown in Fig. 8.24, the current may alternate in direction from wire to wire.

Picket fence →

Magnetic
field lines →

Figure 8.24. Picket fence plasma boundary.

The wires may lie in a plane, they may be coaxial circles on the surface of a cylinder, or they may be straight wires parallel to the axis of a cylinder on whose surface they lie, just to mention three of a number of configurations the picket fence might assume. As long as the separation of the wires is small compared to the radius of curvature of the surface on which they lie, it is appropriate to consider them as straight wires lying in a plane. For such a configuration, at distances of the order of the wire separation or greater, the field will fall as $\exp(-\pi x/a)$ where a is the center to center separation of adjacent wires and x is distance from the plane of the wires. Thus even for strong fields in the plane of the picket fence, the field becomes very small in a distance of, say $3a$ from this plane. The condition for neglect of the field is that the magnetic energy density $B^2/2\mu_0$ be small compared to the plasma pressure nkT. For example, for typical ion source plasmas, with $n = 10^{18}$ m^{-3} and $kT/e = 5$ volts, we find the vacuum B becomes negligible for values less than 1.4 millitesla or 14 gauss.

Ideally, when the picket fence boundary has cusp fields sufficiently strong that electron radii are small compared to the cusp to cusp distance, all electrons are trapped except that fraction which heads for the cusp within some small angle. The reflectivity of the boundary was calculated by Holloday (1954). Such calculations based purely on the motion of a charged particle in a steady magnetic field are valid for the fast electrons in a plasma constrained to one side of such a boundary. Slow electrons can diffuse through the barrier at a much greater rate than one would calculate from single particle trajectories because of the frequency of collisions with other charged particles in the cusp region which tend to fill up the loss cone, and because of the $\mathbf{E} \times \mathbf{B}/B^2$ drifts associated with the fluctuating electric field in the magnetized plasma, as discussed in Sec. 8.3C.

The ions will not be constrained by the magnetic field as effectively as the electrons but the ions are bound to the electrons by space charge considerations which limit their excursion from the plasma to distances of the order of the Debye length, 1.7×10^{-3} cm for a plasma whose density is 10^{12} cm^{-3} and whose electron temperature is 5 eV. We thus see the possibility of creating an approximation to the magic surface of Fig. 8.23.

The multicusped field of the picket fence geometry is most readily produced by arrays of permanent magnets for cusp magnetic fields of less than 0.4 tesla, a very large field for plasmas of density and electron temperature appropriate to ion source plasmas. Various possible arrangements of permanent magnets and iron are shown in Fig. 8.25. Figure 8.25*a* shows rows of linear magnets with alternating directions of magnetization. The resultant field is very similar to that of the picket fence with wires in the positions between the permanent magnet positions. The use of

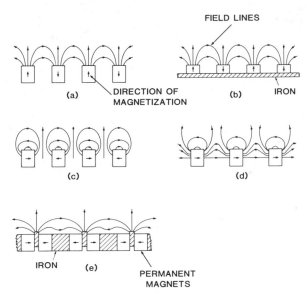

Figure 8.25. Various arrangements of permanent magnets and soft iron for the production of multi-cusped boundaries.

iron, as in Fig. 8.25b, has the advantage not merely of providing a convenient means of holding the magnets in place but of decreasing the required amount of magnet material. Forrester and Busnardo-Neto (1976) show that the soft iron backing is the equivalent of doubling the length of the magnets. Since the cost of some of the highest coercive force magnetic materials can be a major item this is an important practical consideration. In Figs. 8.25c and d the directions of magnetization are parallel to the surface in which the magnets lie but, like the other configurations, yield linear cusp fields resembling those of the picket fence. They have the advantage over the previously described magnets that these arrangements would permit probing the plasma along a field line through the cusp. As far as I am aware, they are not configurations which have been utilized as plasma boundaries. The magnetic fields that would be produced by the arrangements shown in Figs. 8.23 a–d were calculated by Forrester and Busnardo-Neto (1976) on the assumption of uniform magnetization, a valid assumption for very high coercive force magnets such as the rare earth cobalts. The fields they calculate are the vacuum fields. When used as boundaries of a plasma, the fields will be modified by the diamagnetic properties of the medium, but negligibly so near the plane of the cusps for the low β [$=nkT/(B^2/2\mu_0)$] conditions which are always achieved for ion source plasma contained by magnetic boundaries.

The configuration of Fig. 8.25e has the capability of enhancing the field strength at the cusp. For example, working with $SmCo_5$ magnets, which yield cusp fields of 0.35 tesla in the arrangement of Fig. 8.25b, Forrester and Zlotin (1976) measured a field of 0.93 tesla in a configuration resembling that of Fig. 8.25e.

With permanent magnets one can also arrange the magnets to produce point cusps as in the checkerboard pattern used by Limpaecher and MacKenzie (1973; Fig. 8.26). In this arrangement plasma can escape not only into a small angle along each point cusp but, in addition, into a zero field line passing through the center of each local square. Although this boundary has the advantage relative to linear cusps that the field falls off more rapidly with distance from the boundary, the advantage appears to lie with linear cusps and the checkerboard arrangement has not been applied to very many ion sources. (See McAdams et al., 1986, for an example.)

The array of linear cusps can be made to cover the boundaries of an ion source in a variety of arrays such as those indicated in Fig. 8.27. In Fig. 8.27a the magnets are arranged in coaxial circles. It should be noted that this arrangement does not lead to zero magnetic field along the axis. If there is an even number of rings of magnets in the cylindrical surface, their magnetic dipole moments will cancel but the rings on the back face will be left with a net magnetic moment, to the right for the case shown, and a residual field on axis is implied. We shall see that this is deemed desirable for some sources in that it improves the hollow cathode performance.

In Fig. 8.27b the magnets on the cylindrical surface are parallel to the axis and are joined by transverse rows of magnets, as shown. Another arrangement of the rear surface is the "rising sun" configuration shown in Fig. 8.27c. The latter boundary has the symmetry which will lead to effective cancellation of the fields inside the plasma region and provides a convenient central hole which has been utilized for the cathode insertion. Figure 8.27 presents cylinders with flat backs, although several variants are possible. In some sources the rear surface has been conical. In others, as we shall see, the cross section is hexagonal, square, or rectangular.

A view of a plasma boundary with longitudinal magnets, as in Figs. 8.27b and c shows the plasma to have a scalloped boundary with the plasma extending into the cusps. If the plasma were thermalized and the potential constant through the cusp, thermodynamic considerations would imply a uniform density along the field lines right through the cusp region. In fact, the cusps always appear brighter than the main body of the plasma either because of potential gradients through the cusp

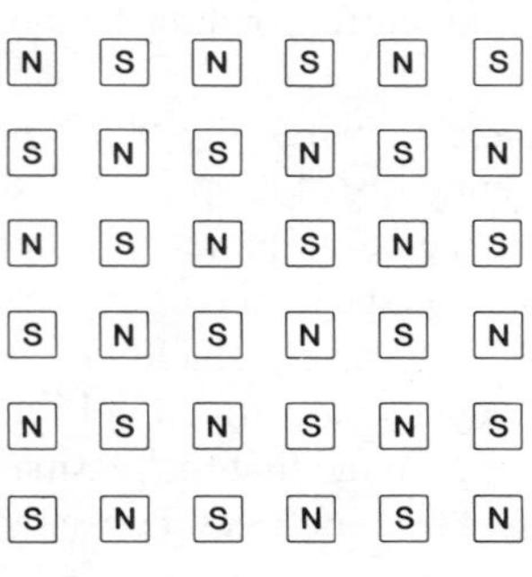

N	S	N	S	N	S
S	N	S	N	S	N
N	S	N	S	N	S
S	N	S	N	S	N
N	S	N	S	N	S
S	N	S	N	S	N

Figure 8.26. End view of permanent magnets in a checkerboard array.

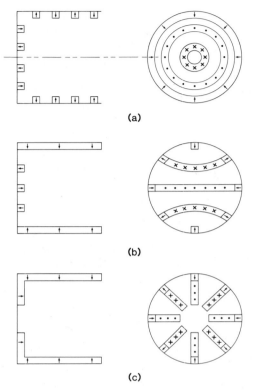

Figure 8.27. Various arrangements of magnets on a cylindrical surface with one open end.

or because ions approach the cusps with directed velocities. I am not aware of a measurement of the density variation along a magnetic field line through a cusp. Hershkowitz et al. (1975) measured the width of the cusp and obtained the value

$$w = 4\sqrt{r_e r_i},\tag{8.6}$$

where r_e is the electron gyroradius corresponding to the velocity $\sqrt{kT_e/m}$, where T_e is the electron temperature and r_i is the ion gyroradius corresponding to the velocity $\sqrt{kT_i/M}$, where T_i is the ion temperature.

When ions are extracted the boundary assumes the shape shown in Fig. 8.28 in which the depth of the scallops alternates between adjacent pairs of cusps. This is due to the $\mathbf{E} \times \mathbf{B}$ effects in the intercusp region. If ions are being extracted, the field, as viewed from the extraction area, will be out of the plane of the figure. In those regions where $\mathbf{E} \times \mathbf{B}$ is outward the scallops will be relatively shallow.

In a source of square cross section with longitudinal cusps, as illustrated in Fig. 8.29a with rows 3.25 cm apart, Goebel and Forrester (1982) found that the plasma was uniform out to 2.75 cm from the cusp plane along a line halfway between

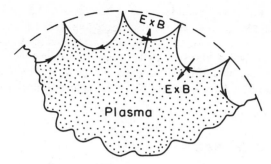

Figure 8.28. View of plasma cusps from the extraction electrode.

cusps where $\mathbf{E} \times \mathbf{B}$ was toward the plane of the cusps, but only out to 3.75 cm from the cusp plane where $\mathbf{E} \times \mathbf{B}$ was toward the plasma. With a source of square cross section Ehlers and Leung (1979) experimented with the two configurations seen in Fig. 8.29. With the configuration of Fig. 8.29b they found that the uniform region of the plasma was too narrow across one diagonal and adopted Fig. 8.29a as the preferred arrangement. The reasons are easy to comprehend. Even the vacuum field could be expected to penetrate further in, as shown by a comparison of field lines in Figs. 8.29a and b. In addition, with the odd number of magnet rows on each face that Ehlers and Leung used, in two of the corners the counterclockwise fields, as viewed from the extraction area, would carry the field penetration inward. With an even number of magnets per face it would be feasible to have the penetrating field clockwise in all four corners. This would improve the situation

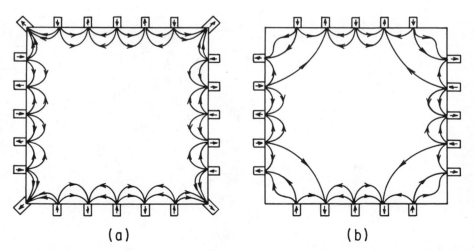

(a)   **(b)**

Figure 8.29. Two possible arrangements of magnets in a square cross section linear cusp containment box.

but, even so, the advantage would appear to be with the configuration in which a row of magnets is placed in each corner of a rectangular cross section source.

It appears that an economy of volume and magnetic field can be effected by building a source in which the spacing between rows of magnets alternated so as to provide the same penetration of the magnetic fields between adjacent pairs of magnet rows. As far as I am aware, such an arrangement has never been utilized.

8.10   Magnetic Materials

It is important to examine the properties of various permanent magnetic materials that have been used or might appear attractive for use in magnetic boundaries. The *B–H* curves in the demagnetization region are shown for five such materials in Fig. 8.30. Ferrimag is the Crucible Magnetics name for their ceramic magnets. Plat-

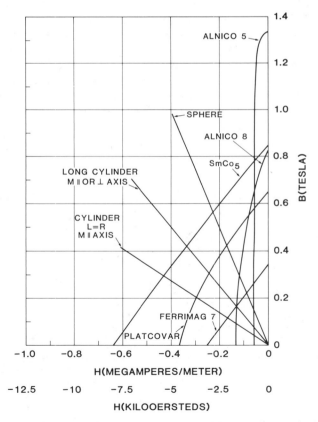

Figure 8.30. *B-H* curves for various permanent magnetic materials in the demagnetization region and *B-H* lines for several geometric configurations.

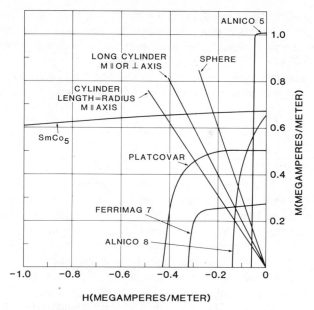

Figure 8.31. Demagnetization curves for various materials and *M-H* lines for several shapes.

covar is a 77% Pt, 23% Co material that is the only one of the listed materials that has the advantages of isotropy, ductility, and machinability.

The magnetic induction B, the field intensity H, and the magnetization or magnetic moment per unit volume M are related by the equation

$$\mathbf{B} = \mu_0(\mathbf{H} + \mathbf{M}),$$

where $\mathbf{B}$ is in tesla, $\mathbf{H}$ and $\mathbf{M}$ are in amperes per meter, and $\mu_0 = 4\pi \times 10^{-7}$ henrys per meter is the permeability of free space. The value of $\mathbf{B}$ at $H = 0$ is a measure of the saturation magnetization, one of the important parameters of a magnetic material. Another is the coercive force, a quantity defined in most textbooks as the value of the reverse field intensity for which the magnetic induction B goes to zero. A much more useful quantity is the coercive force as defined, for example, in the *American Institute of Physics Handbook** as the reverse magnetic field intensity needed to drive M to zero, and we use this definition. The difference between the two definitions of coercive force can be made clear by comparing Fig. 8.30 with Fig. 8.31, a plot of $\mathbf{M} = (\mathbf{B}/\mu_0) - \mathbf{H}$ as a function of $\mathbf{H}$. The usual textbook definition would yield a coercive force for samarium cobalt of 6.4×10^5

*D. E. Gray, ed., *American Institute of Physics Handbook*, 3rd ed., McGraw-Hill, New York, 1972, page 5-3.

A/m, or 8000 oersteds. Figure 8.31 shows this to be a meaningless point. Actually the coercive force for this material is about 1.3×10^6 A/m. The coercive force for the other materials is readily obtained from Fig. 8.31 as the intercept with the abscissa.

If uniform magnetization is assumed, one can find the values of H and B everywhere in the material. This can be done by using equivalent surface currents per unit length

$$\mathbf{J} = \mathbf{M} \times \mathbf{n},$$

where $\mathbf{n}$ is a unit vector normal to the magnet surface or by an alternative approach using a pole strength per unit area equal to the normal component of $\mathbf{M}$. For a sphere the fields are $\mathbf{B} = 2\mu_0\mathbf{M}/3$ and $\mathbf{H} = -\mathbf{M}/3$, uniform throughout the material. This defines the straight line for the sphere shown in Figs. 8.30 and 8.31. The intersection with the curve for a particular material is the actual point that such a sphere would operate when removed from the field that magnetized it. For SmCo$_5$ the value of B in the material would be 0.56 tesla. Since the normal component of $\mathbf{B}$ is continuous, this would also be the field just outside the surface of the sphere at its poles. A material like Alnico 5, which may have some appeal because of its very high saturation magnetization, would give a value of $B = 0.17$ tesla for the same shape. In fact to take advantage of the saturation magnetization of Alnico 5 one would require a horseshoe shape with a small gap to diameter ratio. It is not a good material for a plasma boundary. All of the other materials displayed in Figs. 8.30 and 8.31 do much better.

For a long cylinder uniformly magnetized perpendicular to the axis, the resultant fields, $\mathbf{H} = -\mathbf{M}/2$ and $\mathbf{B} = \mu_0\mathbf{M}/2$, are again uniform throughout the material. For cylindrical magnets uniformly magnetized parallel to the axis the field is not uniform. The B and H values used in drawing the straight lines in the figure are those on axis right at the face, so as to yield the value of B at the cusp. This would be correct for the SmCo$_5$, the Ferrimag 7, and approximately so for the Platcovar, since the intersection is far enough from the coercive force value of H that the magnetization will remain constant through the volume. For the Alnico materials the variation in H through the material would cause a variation in M and the assumption of uniform M would be violated.

For some of these materials, such as the Platcovar and the SmCo$_5$, the cost may be a major consideration. The shape which probably requires the least material for producing line cusps of a given strength is that shown in Fig. 8.32. The effect of the half cylinders plus iron is equivalent to whole cylinders. In Fig. 8.29 we see that a cusp field of 0.42 tesla (4200 gauss) can be obtained with this shape using the SmCo$_5$ material whose B–H characteristic is plotted. The field strength at the surface of the cylinder is independent of the diameter of the cylinder. The ratio of diameter to spacing between adjacent rows can be decreased until the magnetic field between cusps becomes too weak.

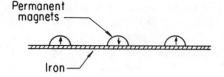

Permanent
  magnets

Iron

Figure 8.32. A magnetic linear cusped boundary
using half cylinders.

8.11   Magnetoelectrostatic Confinement Sources

Normally science and technology advance from the simple to the complex with
each advance depending on prior advances or, as Newton put it, each scientist
standing on the shoulders of those who preceded him. The use of magnetic boundaries for plasma containment failed to follow this precept. The first use of magnetic
boundaries provided by an array of permanent magnets with alternating directions
of magnetization was the boundary of Moore (1969) seen in Fig. 8.33. He called
this boundary the magnetoelectrostatic confinement (MESC) boundary since confinement depended on both electrostatic and magnetic effects. The simpler boundary of Limpaecher and MacKenzie (1973), in which the entire boundary is the
anode of the discharge, came much later, but the developers of that boundary were
unaware of the prior developments. This was a result of the separation between
the electric propulsion community in which the MESC ion sources were developed
and the fusion plasma community in which MacKenzie circulated.

Moore's source, utilizing the boundary of Fig. 8.33, is shown in Fig. 8.34.
The permanent magnets and the iron shell are arranged in hexagonal approximations to coaxial circles. In this source electrons cannot escape through the cusps
since the cusp field lines terminate at cathode potential. The transverse magnetic
fields at the anode strips prevent electrons from reaching this electrode except by
diffusion across the field until they reach a field line which intersects an anode.
Moore estimated that his boundaries reflected 99.95% of the electrons and 92%
of the ions approaching them.

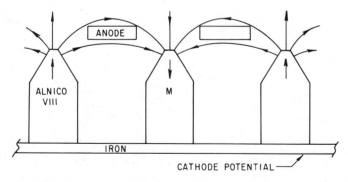

Figure 8.33. MESC plasma boundary of Moore. Arrows indicate the direction of the magnetization
M.

Moore analyzed the performance in terms of rotating plasma streams at the anode surfaces in the $\mathbf{E} \times \mathbf{B}$ directions, alternating in direction from anode to anode because of the alternating direction of $\mathbf{B}$. The plasma velocity at the anode surface is much less than $\mathbf{E} \times \mathbf{B}/B^2$ because of plasma viscosity effects. His entire analysis leading to calculated diffusion currents to the boundaries is not within the scope of our treatment here.

Moore's cathode, which was inserted through a central magnet, as shown, was movable and operation was sensitive to its position.

Moore's boundary appears to have been too effective and he found it necessary to add the plasma anode shown in Fig. 8.34. This anode was small compared to normal anodes and its area so small compared to the extraction area that virtually all ions formed are extracted. The source showed substantially improved performance relative to previous sources. Whereas previous Kaufman-type cesium ion thrusters had achieved an energy input to the ion source of 200 eV/ion at a mass utilization of 97%, this source lowered the required energy to only 100 eV/ion at the same 97% mass utilization. With a different cathode Ramsey (1971) used the same source for mercury where, again, superior performance was demonstrated. Whereas earlier Kaufman thrusters had achieved approximately 190 eV/ion at

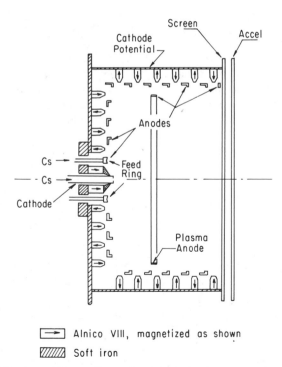

Figure 8.34. MESC source of Moore. Except for the cathode, the cesium vapor feed ring, and the circular plasma anode, the symmetry is hexagonal.

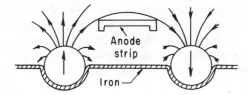

Figure 8.35. MESC boundary of James et al. (1970).

90% mass utilization Ramsey achieved 165 eV/ion at 95%. Extreme uniformity in plasma density was observed in the transverse direction up to a short distance from the boundary where the density fell sharply.

James et al. (1970) made the interesting variant of the MESC boundary shown in Fig. 8.35 and incorporated this boundary into the compact hemispherical shell shown in Fig. 8.36 for a small ion engine for the stationkeeping of a synchronous satellite. It is not a configuration that can be used easily with most permanent magnet materials but the material they used, Platcovar, has the isotropy and machinability that made it feasible. Just as for the Moore source, a plasma anode was found necessary.

MESC sources also have been constructed with rows of magnets parallel to the axis of a cylindrical source, that is, with a magnet arrangement such as shown in Fig. 8.27c. The source of Sovey (1981) was so arranged, except that the rear surface was conical as in the idealized source of Fig. 8.23. Sovey used 20 rows of SmCo$_5$ magnets but only 10 anode rows, and his anodes penetrated sufficiently far in that no separate plasma anode was required.

The arrangement of magnets used by Moore, which was a hexagonal approximation to that displayed in Fig. 8.27a, would not produce zero magnetic field on axis. That of Sovey, as described to this point, would. However, Sovey found zero field undesirable in that his hollow cathode performed much better in a magnetic field. To achieve a field on axis, Sovey removed some of the magnets on the

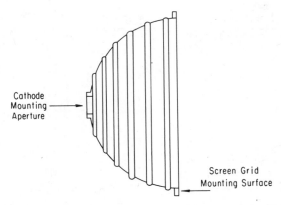

Figure 8.36. Hemispherical shell spun of 0.5 mm iron sheet for the MESC boundary of Fig. 8.35.

conical portion of his chamber for which the magnetization was outward (S poles facing in) while retaining those magnets which were magnetized in the inward direction. This lead to a field at the cathode, diverging strongly toward the screen grid. Optimum performance was obtained for a field at the cathode of 19 mtesla (190 gauss). This optimum would surely depend on the cathode diameter, which for Sovey's source was 6.4 mm.

8.12   MacKenzie Bucket Sources

The magnetic confinement scheme of Limpaecher and MacKenzie (1973) simply placed arrays of point cusps or linear cusps over the surface of a metal container which served as the anode of a discharge. They were concerned with obtaining uniform, dense, quiescent plasmas for basic plasma experiments, but those interested in developing ion sources for the generation of intense neutral beams via charge exchange, previously unaware of developments in the ion engine community, found this containment system provided the solution to the problem of obtaining a uniform current density over the ion extraction area. Sources which utilized multicusped boundaries have taken many forms and have been given many names, among them the MacKenzie bucket source, the name used here.

One such source, that of Forrester et al. (1978), termed IBIS, an acronym for intense boundary ion source, is shown in Fig. 8.37. The magnet rows are longitudinal and the symmetry guarantees zero field on axis and a very weak field over most of the source volume. In intense sources the power loading of the boundaries, as discussed in Sec. 8.1, is such that for heavy ions the magnets can usually be exposed directly to the plasma. This is not possible for light ions such as H^+ or H_2^+ ions, and it is essential to interspace a cooled boundary between the magnets and the plasma to avoid overheating the magnets. For example, the maximum

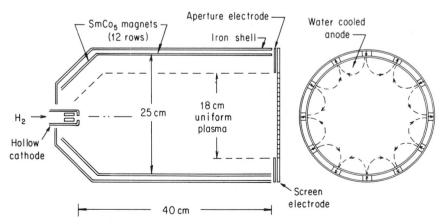

Figure 8.37. IBIS source of Forrester et al. (1978).

temperature before irreversible loss of magnetization sets in is about 200°C for SmCo$_5$. The interposed layer, which can be nonmagnetic stainless steel for low density or water-cooled copper for higher density plasmas, will result in a magnetic field facing the plasma at the cusp which is appreciably less than that at the magnet face unless the magnet dimensions are large compared to the thickness of the interposed metal layer. To obtain a large field at the cusps Forrester et al. used the boundary shown in Fig. 8.38. The iron strips at the face of the magnet rows were brazed into the water-cooled copper liner to give the necessary cusp cooling and concentrate the magnetic flux at the cusp. The field measured at the surface of the iron facing the plasmas was 0.32 tesla, only slightly weaker than the field at the surface of a magnet without the water-cooled liner.

The plasma was found to be uniform in the transverse direction over the central 18 cm, for the 25 cm diameter cooled liner. This defined the ion extraction area. For moderate lengths the lateral flow of ions is so small compared to the flow out the extraction area that the approximation of the plasma as one dimensional should be very good. Under these circumstances the neutral gas density at the exit plane of the ions should be inversely proportional to the length, and this source was given a length of 40 cm, relatively large compared to most source lengths. The large length can also be expected to enhance the ratio of H$^+$ ions to H$_3^+$ and H$_3^+$ ions. The current density extracted from the source was 0.33 A/cm^2 with an arc power consumption of only 330 watts per extracted ampere of ion current. Although the cathode was heated for starting purposes very little power was required to maintain its temperature when the arc voltage was applied. This source was capable of steady state operation.

Even with this source length the effectiveness of the boundary was so great that there was too small an effective anode area. The source started more easily and ran more quietly when the aperture electrode was run at anode potential, providing in effect the plasma anode of Moore. Ideally, it would seem that the best way to provide adequate anode area would be to increase the source length until the ion current to the cusps becomes approximately equal to the beam ion current. As seen in the solution to Problem 8.1, this would probably involve a length of several meters. The advantage of such an oversized source should be a reduced neutral efflux in accord with considerations of Sec. 3.10. A high percentage of atomic ions relative to molecular ions would also be favored by increased source length.

The question of required anode area for stable source operation is one that has

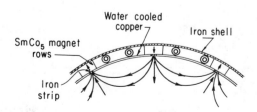

Figure 8.38. Detail of the cusped boundary of the IBIS source of Fig. 8.37.

been attacked by Holmes (1981), Goebel (1982), and Goede and Green (1982). As anode area is decreased the escape of electrons is made more difficult and the plasma potential decreases. Holmes and Goede and Green conclude that unstable operation is achieved when the plasma goes negative relative to the anode. However, the plasma is not an equipotential and its potential decreases along the direction of ion extraction. Goebel and Forrester (1982) found experimentally that stable operation was possible with negative potentials near the extraction area but verified that stable operation appeared impossible when the entire plasma went negative. In the source of Oka and Kuroda (1979) described later, satisfactory operation was associated with a plasma potential near the extraction area of 25 volts negative relative to the anode, with an arc voltage of 90 volts.

Other sources built in my laboratory were of square cross section such as that of Crow et al. (1978) and Goebel and Forrester (1982). The rectangular source of Goebel and Forrester (1981), shown in Fig. 8.39, had an internal anode cross section of 20 cm × 52 cm and provided a uniform plasma over a 10 cm × 40 cm extraction area. Although the magnetic field on the plasma side of the anode was 0.25 tesla, down from the 0.32 value in IBIS, the anode area was still inadequate and stable operation required additional anode area. This was achieved by removing a ring of magnets at the extraction area.

All of the sources described above, except that of Crow et al. which used an oxide coated nickel cathode for an A^+ ion source, utilized hollow cathodes resembling that shown in Fig. 7.19 e. In the large rectangular source Goebel and Forrester used two such cathodes, as shown in Fig. 8.39. There were 30 longitudinal rows of magnets which converged to the two hollow cathodes in side by side rising sun configurations.

Very few of the numerous sources which utilized cusped magnetic boundaries

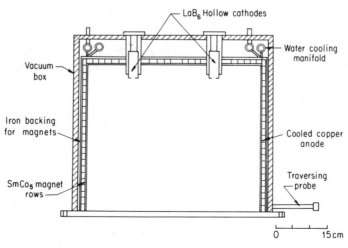

Figure 8.39. Section through the long dimension of the rectangular source of Goebel and Forrester (1981).

covered the anode area as thoroughly as Forrester and Goebel. The cubical source of Ehlers and Leung (1979; Fig. 8.40), for example, had no magnets along the rear side of the box which served as the anode. Their cathode consisted of eight tungsten hairpins, only two of which are seen in Fig. 8.40. Their magnet arrangement was exactly that shown in Fig. 8.29a.

Oka and Kuroda (1979) studied the effect of a multipole boundary on a source which resembled the source of Ehlers et al. seen in Fig. 8.15a, insofar as it was a cylindrical volume in which the cathode was a ring of hairpin peripheral fila-

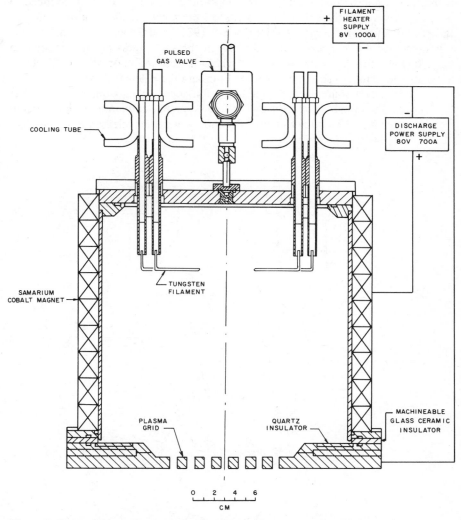

Figure 8.40. Multicusp source of Ehlers and Leung (1979). The cross section is square and the cathode consists of eight 0.15 cm diameter filaments.

ments. They surrounded their source with a ring of electromagnets, as shown in Fig. 8.41, to produce a cusped magnetic field of variable strength, up to 600 gauss, over a portion of the anode surface. As in the source of Ehlers et al., the rear surface and screen electrode were floated and ran at potentials close to the cathode potential to inhibit electrons to these surfaces.

Oka and Kuroda's results included a stabilization of the discharge, a very large lowering of the noise level, an increase in the plasma density, and an improvement in the uniformity all produced by the cusped magnetic field. The plasma potential near the extraction electrode was close to the anode potential for weak magnetic fields, but at approximately 300 gauss, depending on the arc current, dropped to about 25 volts negative, with −90 volts on the cathode, as mentioned above. Based on analyses of Holmes (1981), Goebel (1982), and Goede and Green (1982), it is presumed that the potential at the rear of the source was positive relative to the anode and that this source operated with a large fraction of primary electrons to be able to maintain so large a potential through the plasma.

Another bucket source is the lambdatron ion source of Sakuraba et al. (1981; Fig. 8.42). The unique feature of this source is the backstreaming electron dump whose shape gives the source its name. Secondary electrons released from the accel

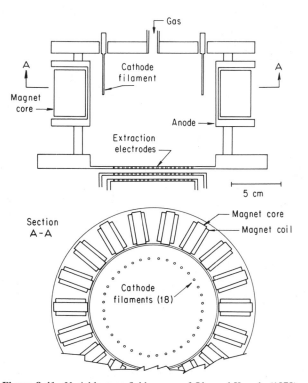

Figure 8.41. Variable cusp field source of Oka and Kuroda (1979).

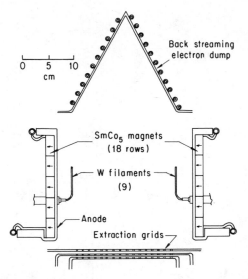

Figure 8.42. Lambdatron ion source of Sakuraba et al. (1981).

electrode by slow ions from the beam plasma or by charge exchange ions generated in the accel gap are accelerated back to the source. For high voltage acceleration the heating produced by these electrons can be substantial. It is this problem that the conical dump of Sakuraba et al. is designed to handle. This source achieved extracted hydrogen ion current densities of 0.27 A/cm^2 with a proton yield of 70%.

When the conical electron dump was replaced with a hemispherical surface covered with linear magnetic cusps the required arc power for a given ion current was reduced to 60–70% of that required for the lambdatron.

8.13   Modified DuoPIGatrons

The use of magnetic cusp boundaries was applied to the duoPIGatron by Stirling et al. (1977) by placing twelve 15 cm long rows of magnets around anode 2 of the source shown in Fig. 8.22, as illustrated in Fig. 8.43. This source then becomes, effectively, a hollow cathode MacKenzie bucket source plus a magnetic field diverging from the region near the cathode aperature to the screen electrode. Such a divergent field was a common feature of the MESC sources but the MacKenzie bucket sources had been built with magnetic arrangements that yielded a magnetic field-free plasma volume.

To produce an ion beam which was uniform over a large rectangular area Menon et al. (1985) built the modification shown in Fig. 8.44 utilizing two LM hollow cathodes. The arc chamber has an internal cross section of 28 cm × 60 cm and is

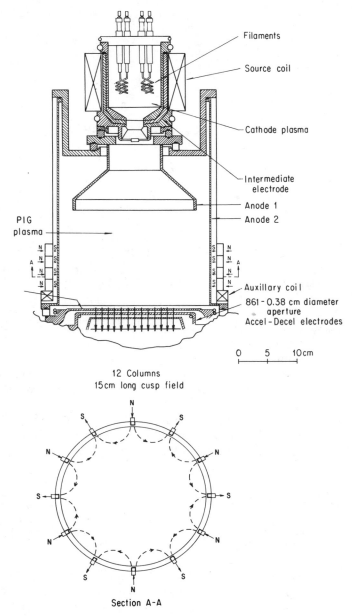

Figure 8.43. DuoPIGatron ion source as modified by 12 rows of magnets.

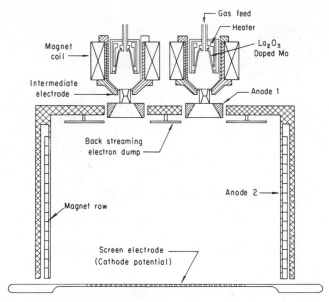

Figure 8.44. Modified duoPIGatron with a 13 cm × 43 cm extraction area.

30 cm long. The plasma density is extremely flat over the 13 cm × 43 cm extraction area. As in the lambdatron, this source incorporates a backstreaming electron dump in its design.

This source was designed for a current density of 0.19 A/cm² of hydrogen ions for 30 second long pulses, although at the time of the referenced paper the current had only been taken to 48 A for a current density of 0.086 A/cm² at the design pulse length. The H⁺ content of the beam was 69.5% at a current density of 0.076 A/cm², and Menon et al. predicted an atomic ion fraction of better than 80% at the goal of 0.19 A/cm². It was not clear what had limited the current density at the time of the writing.

8.14   Magnetic Filtering. The Tandem Cusp Configuration

In hydrogen and deuterium ion sources one of the problems is that of obtaining a large percentage of atomic ions, that is, minimizing the fraction of H_2^+ and H_3^+ ions in the extracted beam. Ehlers and Leung (1981, 1982) discovered that the fraction of H_2^+ ions in the extracted beam was directly correlated to the density of energetic primary electrons in the region of the plasma near the extraction area. To increase the fraction of H^+ ions they introduced the magnetic filter between the rear of the source and the extraction area as shown in Fig. 8.45. This filter was comprised of a very coarse grid of water-cooled ceramic magnets aligned as shown with their magnetization transverse to the ion source. Fast primary electrons

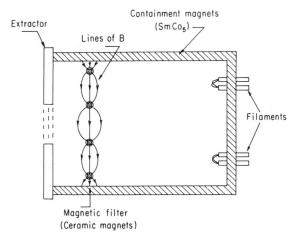

Figure 8.45. Magnetic filter of Ehlers and Leung (1981,1982).

are reflected from this boundary but ions readily move through because of their greater mass, and slow electrons—as always—readily diffuse across the magnetic field, presumably as a result of microinstabilities and collisional effects. The electrons in the volume between the magnetic filter and the extraction plane were observed to have a temperature of only 0.4 eV. This configuration is known as the tandem multicusp configuration. Consistent with previous findings this filter did have the effect of increasing the fraction of H^+ ions in the beam at the expense of the H_2^+ component. It also was observed to produce a flat beam profile out to a larger radius. However, there is a price in lost beam intensity for a given arc current.

By insulating the portion of the source to the left of the magnetic filter in Fig. 8.45, some control over the plasma density in the region near the extractor was obtained. A negative potential on this region increased the plasma density. One can interpret this as the result of a pulling of more positive ions into this region. Slow electrons flow into the region in response to the space charge neutralization requirements of the ions. Conversely, a positive potential on the region to the left of the magnetic filter has the effect of decreasing the plasma potential. In Chapter 10 it will be shown that the ability to create a region with very cold electrons and to control the plasma density with a variable potential are important factors in negative ion generation.

8.15   Periplasmatron

Fumelli and Valckx (1976) and Fumelli and Becherer (1977) developed the interesting configuration shown in Fig. 8.46. In this source the magnetic coils produce fields in opposition to each other so that the resulting magnetic field is the single

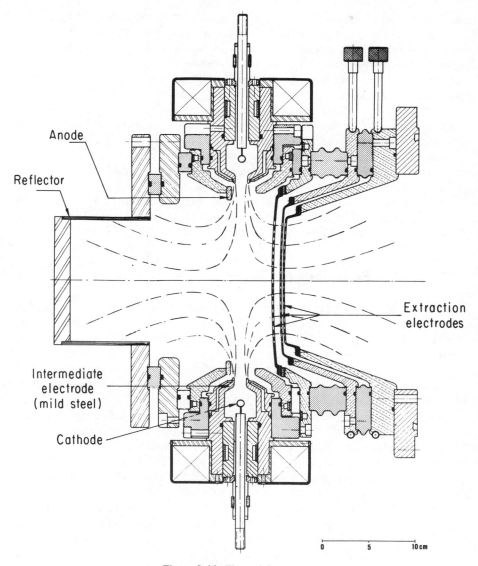

Figure 8.46. The periplasmatron.

ring cusped field indicated by the dashed lines. This source may be viewed as a wraparound duoPIGatron.

The cathode is made of 12 spiral filaments made of 1 mm diameter tungsten wire with a total emitting area of 70 cm^2. Considering the large cathode power required for these filaments it is an advantage that they cannot see and therefore do not heat the extraction grids by their radiation. the intermediate electrode of mild steel which surrounds the emitters is connected to the anode through a 100 ohm resistor and runs close to cathode potential. This electrode plus the tungsten emitters form what is, in essence, an annular hollow cathode filled with a dense plasma. Electrons from this region enter the main plasma region where they are constrained to oscillate along the field lines until they reach the anode. The reflector and the first extraction grid are connected to the anode through 20 ohm resistors and run close to cathode potential. This source has also been constructed in a rectangular version with a 40 cm × 16 cm extraction area with 18 filaments. An interesting feature of this source is that each of the filaments is independently connected to the negative side of the arc supply through a small resistor, a feature which prevents the arc from surging to a single filament and causing that filament to burn through.

The rectangular source has produced a hydrogen ion current of 96 A from a 40 cm × 16 cm area. The required power reported for the circular version was 700–1000 W/A of extracted ions. Gas efficiency is about 60%.

8.16   Radio Frequency Ion Sources

The use of RF for producing an ion source plasma has some important potential advantages. For one thing the need for thermionic cathodes is eliminated. In fact, no internal electrodes may be required. Any arrangement which produces a high frequency electromagnetic field in a region containing a gas or vapor of appropriate density will cause a plasma to form when the field intensity is high enough.

Another advantage accrues from the ease, at microwave frequencies, at least, of efficient coupling of the generated power into the plasma. Even for poor coupling of the electromagnetic wave to an absorptive frequency, reflections can readily be canceled out and, if the plasma cavity has highly reflective walls, then efficient delivery of the electromagnetic power to the plasma must be expected.

An example of an RF positive ion source is that of Delaunay et al. (1983) used in the double charge transfer experiment described in Sec. 10.2. In this source, shown in Fig. 8.47, 8 kW of 8.3 GHz microwaves are delivered by a waveguide to a 5 liter cavity through a ceramic window. A short distance inside the cavity an axial magnetic field drops to 0.28 tesla for which the electron cyclotron frequency is 8.3 GHz and very strong absorption occurs. Although the pressure was only 10^{-3} torr in the cavity, a plasma density of 6×10^{11} cm^{-3} was achieved. At this plasma density the plasma frequency is slightly less than the microwave frequency, so that the microwaves can still penetrate. The plasma expands along the field lines to the single grid extractor used in these experiments. The current density under the conditions described was 150 mA/cm^2, uniform over 10–15 cm in diameter. The ion current was 85% protons.

The number of RF ion sources already pursued or which might fruitfully be investigated is much larger than one but, alas, it will have to go down as a deficiency of this book that RF excited sources are slighted. I slight them not because I feel they are not competitive with DC sources but for other reasons entirely. For

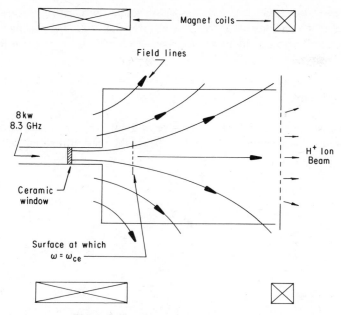

Figure 8.47. Microwave excited ion source.

one thing they have been given very little attention among ion source developers compared to DC sources. For another, they lie outside the domain of subjects with which I have had intimate contact, unlike the other subjects in this volume. The slighting of this area does not imply that I do not believe there is a future for RF sources. In fact I would encourage the venturesome to pursue RF excitation of plasmas as a source of both positive and negative ions.

Problems

Section 8.9

8.1 Assume a cylindrical volume enclosed by 12 linear cusps parallel to the axis. Each cusp has a maximum magnetic field facing the plasma of 0.32 tesla and the enclosed hydrogen plasma has an electron temperature of 8 eV and an ion temperature of 2 eV. For what length would the radial ion escape rate equal the ion current from an extraction aperture 0.15 m in diameter? Assume H_2^+ ions.

8.2 Assume the ion source of Problem 8.1 is 2 m long and that the neutral molecule density is constant over the entire length. Suppose that the rear of the source is unmagnetized so that half of the ions go to the rear and half toward the extraction area. Take the ionization to be produced by 50 eV primary electrons which constitute 5% of the electron density in the plasma. The cross section for ionization by these electrons is 9×10^{-17} cm^2. What is the neutral gas density?

General Problem

8.3 An ion source fed with H_2 at the rate of 1 torr L/s (torr-liters per second) generates a 1 A beam of H^+ ions. What is the mass utilization efficiency?

Chapter 9

Surface Ionization Sources

9.1 Description of Surface Ionization

When an atom of a gas or vapor impinges on a metal surface and then evaporates, energy considerations, in certain cases, favor the abandonment by the atom of an electron, that is, the evaporation as a positive ion. This phenomenon has been used as the basis of intense Cs^+ ion sources with application to other ions as well.

The requirement, illustrated in Fig. 9.1, is that the ionization potential I be less than the work function ϕ. In that case the system is in its lowest energy state when the most lightly bound electron in the atom drops into the metal. A formula for the relative numbers of atoms and ions evaporating can be derived readily. When the atom is close enough to the metal so that the outermost electron can, by tunneling, equilibrate with the electrons in the metal it shares the Fermi energy level corresponding to the top of the continuum of filled levels.

In this case the probability of finding the atomic level occupied is given by the Fermi–Dirac function

$$P_a = \left(1 + \exp \frac{\phi - I}{kT/e}\right)^{-1},\qquad(9.1)$$

since the level is at an energy $e(\phi - I)$ above the Fermi level. P_a represents the probability of the atom evaporating as a neutral and $(1 - P_a)$ or

$$P_p = \left(\exp \frac{\phi - I}{kT/e}\right)\left(1 + \exp \frac{\phi - I}{kT/e}\right)^{-1}\qquad(9.2)$$

229

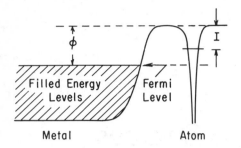

Figure 9.1. Electron energy levels for an atom near a metal surface.

represents the probability of the atom escaping as a positive ion. The ratio P_p/P_a yields

$$\nu_p/\nu_a = \exp\left[e(\phi - I)/kT\right], \tag{9.3}$$

where ν_p and ν_a represent the number per unit area per unit time of evaporating ions and atoms, respectively. Neglected in this derivation are the weighting factors

$$g = 2S + 1 \tag{9.4}$$

for the two states. With these Eq. (9.3) becomes

$$\nu_p/\nu_a = (g_p/g_a) \exp\left[e(\phi - I)/kT\right]. \tag{9.5}$$

Usually the evaporating species is an alkali atom for which $g_p = 1$ and $g_a = 2$.

Equation (9.5) is known as the Saha–Langmuir equation because it was derived by Langmuir on the basis of the Saha equation. I find the method given here more direct.

For cesium evaporating from tungsten $I = 3.87$ volts and $\phi = 4.62$ volts and with T taken as, say, 1500 K, Eq. (9.5) leads to $\nu_p/\nu_a = 165$. If $T = 1000$ K is used, then we obtain $\nu_p/\nu_a = 3000$. Although low temperature favors a high ion yield, the fraction of ions in the efflux from the surface is very high as long as ($\phi - I$) $\gg kT/e$. The minimum temperature that can be used is that which will maintain the tungsten surface free of an adsorbed layer of cesium, which has the effect of lowering ϕ. To understand and explain the observations concerning surface ionization it is necessary to discuss this adsorbed layer.

9.2 Surface Adsorption

The phenomenon associated with the evaporation of cesium from a tungsten surface was widely studied, for example, by Langmuir and Kingdon (1925), Copley and Phipps (1935), and Datz and Taylor (1956), but the definitive work is that of

Taylor and Langmuir (1933). (However, note the reference to Papageorgopoulos and Chen in Sec. 9.4.) Taylor and Langmuir took data over a range of 400–1300 K but experiments of Shelton et al. (1959) justify extrapolation of their results to at least 1500 K.

At any given rate of delivery of cesium to a heated tungsten surface there will be an equilibrium adsorbed surface density of cesium atoms. If the surface is hot enough, the adsorbed layer will be a sufficiently small fraction of a monolayer that the work function will be close to that of pure tungsten, but even a small percentage of a monolayer causes a sufficient decrease in the work function that the fractional ionization may be severely degraded.

By a variety of very careful measurements Taylor and Langmuir were able to determine the number of cesium atoms in a filled monolayer (3.56×10^{14} cm^{-2}) and the fraction θ of this density which obtained for any cesium delivery rate and tungsten temperature. They were then able to relate the work function to θ and obtain the ion and atom evaporation rates and electron emission as a function of θ and T. Their results, as presented by Forrester (1965), are as follows. The atom and ion emission rates, ν_a and ν_p, respectively, in number per square centimeter per second, may each be written as

$$\nu = \exp\left(A - B/T\right), \tag{9.6}$$

where T is the absolute temperature and A and B are functions of θ, different, of course, for atoms and ions. For atoms

$$A_a = 61 + 4.8\left(\theta - \frac{1}{2}\theta^2\right)$$

$$+ \ln\frac{\theta}{(1-\theta)} + \frac{1}{(1-\theta)} \tag{9.7}$$

and

$$B_a = 32{,}380/(1 + 0.714\theta). \tag{9.8}$$

For ion emission

$$A_p = A_a - \ln 2 \tag{9.9}$$

and

$$B_p = B_a - 8681 + 11{,}606V_c \tag{9.10}$$

where V_c is the contact potential between pure tungsten and the cesium coated surface, the amount by which the cesium coating lowers the tungsten work function. V_c was measured and tabulated by Taylor and Langmuir. Forrester (1965)

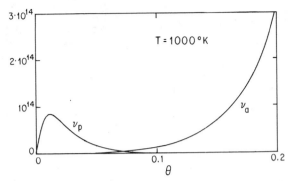

Figure 9.2. Zero field ion (ν_p) and atom (ν_a) evaporation rates in particles per square centimeter per second as a function of surface coverage at 1000 K.

made a least-mean-squares fit between their data and a four term polynomial and showed that the contact potential could be represented by

$$V_c = 10.679\theta - 22.982\theta^2 + 42.53\theta^3 - 34.91\theta^4 \qquad (9.11)$$

to well within error limits.

Isothermal curves of evaporation rates as a function of θ, such as that shown in Fig. 9.2, may be plotted from Eqs. (9.6)–(9.11). Such curves show a more complicated behavior than might have been expected. The ion efflux rate rises to a maximum at a value of θ approximately 0.01 and then falls sharply. The atom efflux rises steadily with θ but is so small compared to the ion efflux at coverage close to 0.01 that the total evaporation rate undergoes a maximum at essentially the same value of θ at which the ion current maximizes.

This implies a discontinuity in ion current density from a tungsten surface as a function of delivery rate. For example, imagine delivering cesium vapor to a 1000 K tungsten surface at a rate slowly increasing from zero to 8.5×10^{13} atoms/cm²-sec. During this process θ would increase to about 0.012 and the evaporating cesium would be virtually 100% ionized. A further slight increase in the delivery rate could not be matched by a small increase in θ. Rather θ would necessarily jump to a value of about 0.156 where the evaporation would be almost exclusively in the form of neutral atoms.

It will be useful to note that in the limit of very small θ, Eqs. (9.6)–(9.11) yield

$$\nu_p = \theta \exp\left(61.31 - 23{,}699/T\right). \qquad (9.12)$$

9.3   Ion Current Density

While the parameter θ is helpful in understanding surface ionization phenomena, Fig. 9.3, in which the ion current density $J_p = e\nu_p$ is plotted as a function of temperature with the fraction of neutrals as a parameter, is a more instructive guide

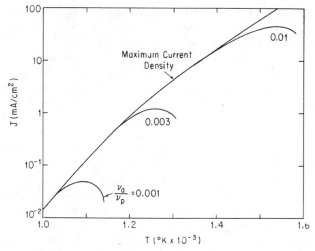

Figure 9.3. Zero field Cs$^+$ ion current density from a tungsten surface as a function of temperature for various atom to ion efflux ratios.

to the usefulness of surface ionization as the basis of high current density ion sources. We see that the available current density increases rapidly with temperature and reaches interesting current densities at accessible temperatures. For example, at 1300 K, a very modest temperature for tungsten, Fig. 9.3 predicts an available current density of over 4 mA/cm^2 with less than 1% neutral efflux. Note that this is a large current density for so heavy an ion as cesium. For orientation it is the maximum current density which the Child equation would permit to flow between parallel planes 1 cm apart with a potential difference of 9 kV. Or, to compare with plasma sources, 4 mA/cm^2 is the ion current which Eq. (3.69) would yield as the ion current density available from a 1 eV cesium plasma with a density of 6 × 10^{11} electrons/cm^3.

As temperature increases it might be noted that the fraction of neutrals in the beam increases but remains at such low levels that for most purposes we have virtually 100% ionized emission. Specifically we see that we can get up to 50 mA/cm^2 at 99% ionization.

9.4 Electron Emission

Although this chapter is concerned primarily with ion emission, this appears to be the appropriate place to discuss the important subject of electron emissions from cesium coated tungsten surfaces. Normally the cooling of an electron emitter results in a monotonic decrease in electron emission satisfying the Richardson–Dushman equation

$$J = AT^2 \exp\left(-e\phi/kT\right). \tag{2.63}$$

However, if an emitter such as tungsten is in a cesium environment, the adsorption of cesium so lowers the work function that there is a range of temperatures over which decreasing temperature yields increasing emission, leading to the S-shaped curves of Fig. 9.4. In this figure, taken directly from Taylor and Langmuir (1933), μ_a represents the rate in atoms per square centimeter per second at which atoms are delivered to the surface. The temperature on each curve is the cesium reservoir temperature that will yield the given value of μ_a. The ordinate is $\log_{10} \nu_e$ and the electron current density in amperes per centimeter square per second can be obtained from ν_e by multiplying by the electron charge 1.6×10^{-19} coulomb.

The diagonal straight lines intersecting the emission curves give the values of θ, the adsorbed monolayer fraction. For example, consider the curve for $\mu_a = 10^{15}/\text{cm}^2$-s. At temperatures greater than 1400 K the surface would be so clear of adsorbed cesium that the emission would be that of pure tungsten. When the temperature drops to 1250 K, θ has increased to 5% of a monolayer and the emission is starting to rise. Even though the current has not significantly risen at this point, the emission is approximately 100 times that of pure tungsten. As the temperature

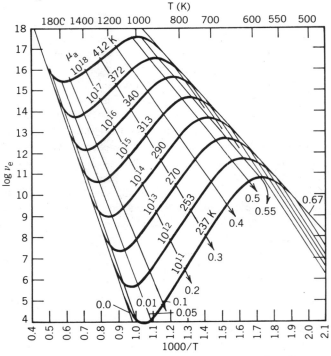

Figure 9.4. Field-free electron emission from tungsten in equilibrium with cesium vapor. The rate of arrival (μ_a) of Cs and the Cs bath temperature are given on each curve. Note that the ordinate is log to the base 10 and that the ν_e and μ_a are in square centimeters per second. Diagonal straight lines give the adsorbed Cs density in fraction of a monolayer. From Taylor and Langmuir (1933).

drops to 750 K, where θ has risen to 0.55, the current is greater by a factor of 10^4 than it was at 1250 K and of the order of 10^{15} times greater than the emission one could predict for tungsten at 750 K.

Although the current drops as the temperature is decreased further, the work function continues to decrease as θ increases, reaching a minimum value of $\phi = 1.72$ volts at $\theta = 0.67$. The constant A for the $\theta = 0.67$ line is 186 A/cm^2 $- $ K^2. More recent and probably more exacting studies of the cesium–tungsten system by Papageorgopoulos and Chen (1972, 1973) have yielded values of $\phi_{min} = 1.56$ volts at $\theta = 0.5$ for the (112) crystal surface and $\phi_{min} = 1.60$ volts at $\theta = 0.25$ for the (100) crystal surface.

It is worth noting that the currents shown in Fig. 9.4 are corrected to zero field. In a manner characteristic of composite surfaces, as discussed in Sec. 7.3, the emission shows a very large Schottky effect, especially so in the vicinity of $\theta = 0.6$, and even modest fields will produce emission currents five times as large as those at the peak of the S curves of Fig. 9.4.

9.5   A Possible Front Feed Ion Source

The available ion current density shown in Fig. 9.3 suggests the possibility of an ion source such as that shown in Fig. 9.5. In that figure cesium vapor is shown delivered to the contoured face of a heated tungsten plate through fine holes on slits in a tubular accel electrode structure. Ions formed at the tungsten surface are shown to be formed into an array of beams by the accel–decel electrode system. The concept is simple but presents the following significant problems:

1. *Neutral cesium vapor distribution.* The delivery tubes are shown at 600 K in order to get an adequate flow of cesium without any condensation in the tubes.

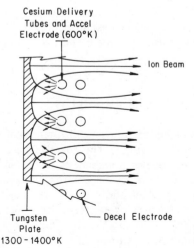

Figure 9.5. Conceptual front feed surface ionization ion source.

It will be seen in Sec. 9.6 that at such temperatures an adsorbed layer of cesium will uniformly cover the tube so that vaporation will take place in all directions, delivering as much cesium downstream as toward the ionizer plate. This problem probably can be avoided by running these tubes much hotter, at temperatures close to the ionizer temperature, where, as we shall see, the diffusion length is very small.

2. *Electron emission from the accel electrode.* The heated cesium coated electrode will be a copious source of electrons. These electrons might be used to heat the ionizer but this makes for an undesirable lack of independence of the source parameters. For example, the ionizer temperature will depend on the cesium delivery rate and accel temperature. It might be a manageable system but it presents difficulties.

3. *Charge transfer in the electrode region.* Typically one might imagine an emission current density from the ionizer of about $J = 2 \text{ mA}/\text{cm}^2 = 20 \text{ A}/\text{m}^2$. At a delivery temperature of 1160 K (0.1 eV) the atomic velocity, $v = \sqrt{kT/M} = 265 \text{ m}/\text{s}$ leads to a neutral atom density in front of the ionizer of

$$n_a = J/ev = 4.7 \times 10^{17} \text{ m}^{-3} = 4.7 \times 10^{11} \text{ cm}^{-3}.$$

Although the charge transfer cross section σ_c varies with ion energy, one can see from Fig. 18 of Mapleton (1972, p. 211) that from 10^2 to 10^4 eV, $\sigma_c = 3 \times 10^{-14}$ cm^2 is a fair approximation. If the acceleration gap were, say, $l = 0.5$ cm, the fraction of ions which exchange an electron with the incoming neutrals, $\sigma_c n_a$, is about 7×10^{-3}. This is not a very large fraction but the parasitics formed, as described in Sec. 5.6, can be damaging. Many of the slow ions formed in the electrode region will be accelerated into the electrodes, where the effects of sputtering limit electrode lifetimes.

As far as I am aware, these reasons have precluded the development of sources which can be typified by Fig. 9.5 in favor of sources in which cesium is fed through a porous tungsten aggregate. In this case the delivery of the cesium is largely by surface diffusion and it is necessary to consider that aspect of the cesium–tungsten system.

9.6 Surface Diffusion Rates and Characteristic Lengths

When cesium is fed to a metal surface the adsorbed cesium layer will spread over the surface as the surface evaporates cesium ions and atoms. The effect is especially well displayed with the ion emission microscope of Marchant et al. (1963) and Forrester et al. (1963). In that instrument ions from an ion emitting surface are accelerated through an electrostatic lens system which forms an image, magnified by a factor of 25–200, on a 400 line per centimeter nickel screen. Secondary electrons from the screen are then accelerated through a 10 kV to an aluminized phosphor screen which gives a bright image of the emission pattern.

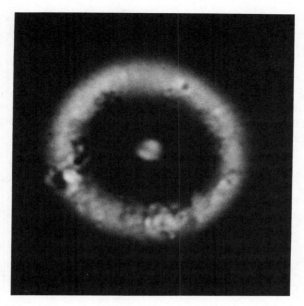

Figure 9.6. Ion emission micrograph of the face of a heated tungsten plate with a rear-fed 25–50 μm diameter hole. (Taken in 1963 or 1964 by B. Marchant, G. Kuskevics, and A. T. Forrester; previously unpublished.)

Figure 9.6 is a particularly simple display of the effects of surface diffusion obtained with the ion emission microscope. In this experiment cesium is rear fed through an approximately 25–50 μm diameter hole in a heated tungsten plate. Although the hole appears as the small bright area in the center, the surface near the hole and out to several diameters from the hole is dark because coverage θ is too large, that is, it is in the neutral atom emitting region of Fig. 9.2. Under the influence of outward diffusion and atom emission the value θ decreases until a value of a few percent is reached when the ion emission becomes important. The brightness increases as θ decreases to about 0.01 and then decreases rapidly as θ falls toward zero. The somewhat mottled appearance of the ring is attributed to the multicrystalline nature of the tungsten sheet. As the cesium feed was decreased the bright ring was observed to move inward until it disappears into the hole.

It is somewhat puzzling that the hole in Fig. 9.6 appears bright when the front surface of the ionizer adjacent to the hole is nonemitting, since the surface coverage is in the hole must be greater than on the surface. In an unpublished 1964 Electro-Optical Systems report I attributed this to the variation of the work function within the hole. The resultant variation in contact potential can be shown to produce electrostatic fields which gather ions from the cylindrical walls of the hole and accelerate them to the emission aperture.

Notwithstanding such effects, which can enhance the ion emission from pores, it appears that the required fineness of a rear-fed porous ionizer material will be determined essentially by the diffusion length, the distance in which the surface coverage will drop from a maximally emissive value to $1/e$ that value on a plane surface.

The flux of cesium adatoms per unit length which flows from more densely to less densely covered portions of a surface can be expressed by $D\nabla\sigma = -D\sigma_0\nabla\theta$ where $\nabla\sigma$ is the gradient of the surface density σ, D is the surface diffusion coefficient, or diffusivity, and $\theta = \sigma/\sigma_0$ is the fractional surface coverage. The condition that the net rate at which adatoms that flow onto each area of the surface exactly balance the evaporation rate $\nu(\theta)$ is then

$$\sigma_0 D\nabla^2\theta = \nu(\theta)$$

or, for a one dimensional situation,

$$\sigma_0 D d^2\theta/dx^2 = \nu(\theta). \tag{9.13}$$

At very low coverage $\nu(\theta)$ can be taken as linear leading to

$$(d^2\theta/dx^2) - \left[\nu'(0)/\sigma_0 D\right]\theta = 0, \tag{9.14}$$

where $\nu'(0)$ is the low coverage limit of $d\nu/d\theta$. The solution is given by

$$\theta = \theta_0 \exp\left(-x/\delta\right), \tag{9.15}$$

where the diffusion length δ is given by

$$\delta = \left[\sigma_0 D/\nu'(0)\right]^{1/2}. \tag{9.16}$$

This can also be expressed as

$$\delta = \sqrt{D\tau}, \tag{9.17}$$

where $\tau = \sigma_0/\nu'(0)$ is the desorption time.

The value of D have been measured by Langmuir and Taylor (1932) over the temperature range from 650 to 812 K, where they found that it could be represented by

$$\log_{10} D = -0.70 - 3060/T. \tag{9.18}$$

However, Forrester (1965) showed that it is not possible to use Eq. (9.18) as the basis for an extrapolation to the temperature domain of interest for high current ion sources, that is, 1300–1500 K. Not only is this procedure dangerous because it goes 600 K beyond the measurement region, but the diffusion coefficient in the range of 1300–1500 K is limited by a process completely different from that for the region below 899 K, as pointed out by Zuccaro et al. (1960).

In the lower temperature region the magnitude of D is limited by the average time which an adatom spends at a trapping site. The formulation of Eq. (9.18) is correct for a process of this type limited by an activation energy. Time spent by an adatom in transit from one site to another apparently is negligible compared to trapping times in the temperature range < 812 K.

If transit times between collisions of cesium adatoms with tungsten atoms limits the rate of diffusion, it can be shown that the surface diffusion constant will be given by

$$D = \bar{v}\lambda/4, \tag{9.19}$$

where $\bar{v}$ is the mean speed and λ the mean free path. If we take the mean free path as the mean distance between atoms in one crystal face, say the (110) face, we get $\lambda = 2.65 \times 10^{-8}$ cm, and with the mean speed of a two dimensional gas given by $\sqrt{\pi kT/2m}$ we get the migration time limited diffusion constant seen in Fig. 9.7. With both trapping and migration effective, the resultant D, obtained by combining the two D's as $D_1 D_2/(D_1 + D_2)$, is close to the migration limited diffusion constant in the range of 1300–1500°K, where high current density sources must operate.

We now have D as a function of temperature. The surface coverage σ_0, needed to obtain δ from Eq. 9.16, was determined by Taylor and Langmuir (1933) to be

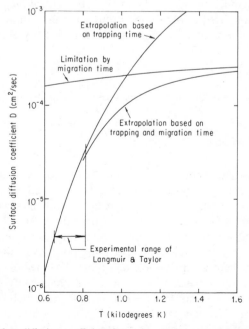

Figure 9.7. Surface diffusion coefficient for cesium on tungsten at low concentrations.

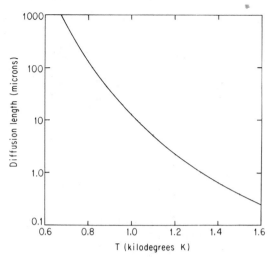

Figure 9.8. Diffusion length for cesium on tungsten at low surface coverage as a function of temperature.

3.6×10^{14} cm^{-2} for a microscopically smooth tungsten surface and 4.8×10^{14} cm^{-2} if the apparent area of a normal smooth tungsten surface is used. Considering the uncertainties in our knowledge of D, the question of the appropriate value to use can easily be avoided by taking a value in between, say, $\sigma_0 = 4 \times 10^{14}$ cm^{-2}.

All that is still required to obtain the diffusion length δ then is $\nu'(0)$. At low coverage ν is virtually identical with ν_p so that Eq. (9.12) yields

$$\nu'(0) = \exp\left(61.31 - 23{,}699/T\right). \tag{9.20}$$

The diffusion length as a function of T is shown in Fig. 9.8.

9.7   Porous Tungsten Ionizers

It seems clear from simple considerations that an effective porous tungsten ionizer needs to have hole sizes and distance between holes which are of the order of the diffusion length δ or, if possible, smaller. If the hole diameter is large compared to δ and the end of the hole has a coverage near the 0.01 value of θ for which maximum emission occurs, then the aperture facing the acceleration electrodes would present a large solid angle to deeper lying portions of the channel where $\theta \gg 0.01$ and a large neutral efflux could be expected. If the hole is small compared to δ but the distance between holes is large, then presumably one could operate satisfactorily with ion emission from the holes and from a small region surrounding the holes but power efficiency would be somewhat degraded by having areas of the surface which are radiating but not emitting.

Porous materials used for this purpose have usually been made by pressing and sintering fine tungsten powders, but those attempting detailed calculations (e.g., Zuccaro et al., 1960; Nazarian and Shelton, 1960) have found it necessary to confine their attention to straight cylindrical pores. One such calculation by Forrester (1965) finds the solution within a pore, where transport is both by vapor flow and surface diffusion and matches this solution to the front face, where transport is strictly by diffusion. Forrester found that the system is characterized by various states with critical transitions among these states. Slight changes in ionizer temperature or cesium vapor feed pressure discontinuously change the system from a pore emitting state to a front surface emitting state or a nonemitting state.

This phenomenon was observed with the ion emission microscope previously described using ionizers made by the parallel stacking of 12-μm diameter tungsten wires. Figure 9.9 is an emission micrograph of a situation in which three states have been made to coexist.

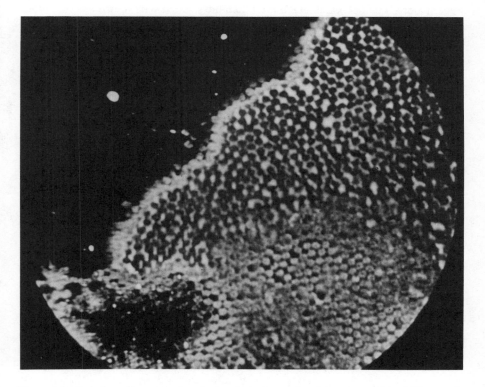

Figure 9.9. Emission micrograph of a wire bundle ionizer under triple state conditions. The area in the upper right is emitting ions from the pores, with the front surface having too little cesium coverage to yield noticeable ion current. In the lower right, the wire ends are emitting ions, and only neutral cesium emerges from the pores. In the upper left the ionizer is in a flooded state and only neutrals are being emitted. The bright band between the flooded area and the area in which the wire ends are impoverished is caused by diffusion over the exposed surfaces.

A more complete calculation was made by Bates and Forrester (1967), who included in addition to vapor transport and surface diffusion the effects of fields in the capillary due to variations of the contact potential. Although these fields have the effect of increasing the ion current density from the pores, in general, the results of Forrester (1965) were shown to be fairly accurate. An important conclusion is that it seems generally true that the greatest average ion current density is obtained when emission is primarily from pores rather than from the front faces.

The wire bundle ionizers referred to above and used in obtaining Fig. 9.9 were made by stacking fine wires inside a small tube which was subsequently shrunk in diameter by swaging. The interstices were then filled with copper for convenient machining to flat surfaces. The copper could then be evaporated out and the disk mounted on an assembly which could be heated and fed with cesium.

Materials used as the ionizers in practical high current density ion sources, however, were generally made not by the parallel stacking of tungsten wires but rather by the sintering of very fine powders to 80–83% of solid tungsten density. Such materials can be made much finer than the stacked wire ionizers and in much larger sizes. The best ionizer materials were made by starting with 2–10 μm diameter spherical powders and pressing and sintering to the appropriate density. The sintering temperature was about 1800 K, sufficiently high compared to the approximately 1400 K operating temperature so that these materials proved extremely stable in operation.

9.8   Ion Source Configurations

Large efficient ion sources using rear-fed porous tungsten ionizers have been constructed and operated continuously for thousands of hours. If we consider only the porous ionizer, then the sources are of two types, one utilizing an array of small porous tungsten disks, which is called the button source, and another utilizing a large multiply contoured surface and termed the sastrugi source. Figure 9.10 is a schematic of the construction of a button source of a type developed at Electro-Optical Systems for space propulsion. The construction is fairly complicated. The porous buttons are hermetically sealed in a close packed array of recessed holes in a molybdenum plate. This alone presents a challenge. Brazes were developed which could do the task, but since capillary action tends to draw brazing material into the pores it was necessary to develop a special material that would alloy with the tungsten before any deep penetration occurred. Ultimately it proved superior to electron beam weld each button into place, but this requires a skilled electron beam welder with high quality equipment.

The need for uniform heating of such an extended array dictated that the rear of the vapor manifold, which contained the heater element, be connected to the front by a multiplicity of conduction paths as shown. These were the triangular pieces left in the crucible when three sets of parallel end mill grooves were run in the three directions which followed the rows of button holes in the button mounting plate. The molybdenum plate with the ionizer buttons must then be brazed to the

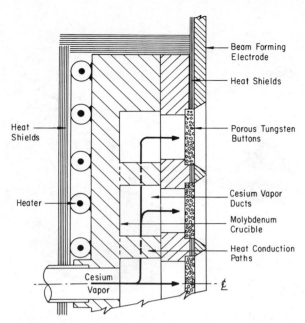

Figure 9.10. Button ion source construction schematic.

crucible in such a way that the peripheral seam is vacuum tight and with each heat path forming a dependable contact.

Efficient heat shielding could not readily be accomplished with an assembly of this kind heated to 1400 K by radiant heaters. Good thermal efficiency requires swaged heaters brazed to the back of the crucible, as illustrated. In these heaters a wire, for example, nichrome or tantalum depending on the desired temperature range, strung with beads of a high temperature insulator, such as alumina or magnesia, is inserted into a sheath made, for example, of stainless steel or molybdenum. The sheath is then swaged down to, say, 3 mm in diameter. The densely packed alumina or magnesia has sufficiently high thermal conductivity that the heater wire need not run much hotter than the sheath and, with well designed terminations, these heaters can have a very extended lifetime. The fact that this heating system exposes no parts which are significantly higher than the ionizer temperature simplifies the heat shielding problem and improves the energy efficiency of the source. For temperature up to 1500 K there exists a multiplicity of ceramic fiber materials such as Fibrefrax and Zircar, which can serve as effective heat shielding for the rear side of the assembly.

Finally, the front side of this assembly requires a beam forming electrode which can be, as shown, thermally insulated from the ionizer assembly.

In Fig. 9.10 the thickness is exaggerated relative to the diameter. Figure 9.11 shows an actual ion source assembly without the insulation and beam forming electrode.

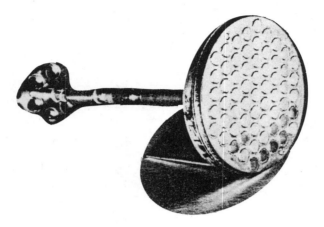

Figure 9.11. Photograph of 61 button ion source with beam forming electrode and heat shielding removed.

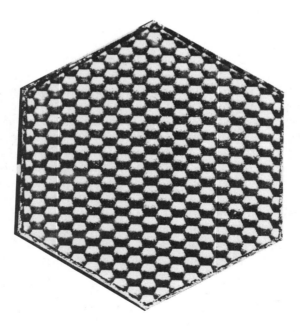

Figure 9.12. Sastrugi ionizer front surface showing hexagonal pattern of sharp ridges. From Seitz and Eilenberg, 1967.

Figure 9.13. Sastrugi ionizers in various stages of assembly. From Ernstene et al., *Journal of Spacecraft and Rockets*, **3**(5), 744 (1966), Figure 1.

In the alternative approach a large sheet of porous tungsten has spherical depressions machined onto its surface in a close packed hexagonal array. If these depressions merge into each other along sharp ridges, a pattern such as shown in Fig. 9.12 is obtained. As shown in Fig. 9.5 almost all of the ions will focus through the apertures, but in this case the beamlets will have cross sections which are

Figure 9.14. Assembled ion source utilizing the Fig. 9.13 ionizer assembly. Ernstene et al., *Journal of Spacecraft and Rockets*, **3**(5), 744 (1966), Figure 3.

approximately circular. This ionizer configuration has been termed a sastrugi ion-izer. Figure 9.13 shows sastrugi ionizers in several stages of assembly and Fig. 9.14 shows the total ion source assembly.

In the assembly of a sastrugi ionizer it is necessary to have the heat conduction paths actually machined into the rear side of the porous tungsten, rather than in the crucible, as for the button ionizer. In both types of source the accel electrodes need to be made of copper. In that case the copper which is sputtered onto the surface reevaporates harmlessly. Materials such as molybdenum, nickel, and steel deposit materials which ultimately clog the pores in the ionizer. Sources of both types have operated satisfactorily with current densities of a few milliamperes per square centimeter, which was as much as they were called upon to deliver.

9.9   Alternative Ions and Ionizers

We have dealt, to this point, entirely with the cesium–tungsten system. In fact several ionizer materials are possible and ions of many species can be efficiently made. Perel (1968) reported on the production of small beams of Rb^+ and K^+ as well as Cs^+ ions from porous tungsten ionizers.

Equation (9.3) indicates that what is required for a high percentage yield of positive ions is a high work function ionizer and there are higher work function materials than tungsten. For this purpose Wilson (1966) measured the work func-tion ϕ of various metals and showed that the work functions of polycrystalline rhenium, osmium, iridium, and platinum were 4.96, 4.83, 5.27, and 5.7 volts, respectively, all substantially higher than that of tungsten. Furthermore, composite surfaces, such as that produced by an adsorbed layer of oxygen on tungsten, can produce values of the work function of the order of 10 volts. However, as covered in Secs. 9.2 and 9.3, the production of high current density involves a dynamic state in which the ionizer material is coated with an adsorbed fraction of monolayer of the material to be ionized. In fact the value of the work function in itself tells only the limiting ionization efficiency as the current density approaches zero and nothing about the maximum current density. Tests of cesium on iridium, for ex-ample, have shown that at a given temperature the maximum current density is much less than for cesium on tungsten.

However, the higher work function substrates do open the way to the formation of ions of species other than cesium. For example, on a surface such as iridium one can calculate, in the limit of low delivery rates, highly efficient ionization of rubidium and potassium, efficiencies of the order of 50% for sodium and barium and 25% for lithium. In fact, Wilson (1973) has made indium and gallium ion sources utilizing an iridium ionizer, but only very small currents and low percent-age ionization. Daley et al. (1966) made an indium ion source in which the req-uisite high work function was achieved by constantly feeding oxygen to the emit-ting face of a rear-fed porous tungsten button. This was a successful source but it might not be feasible to extend this technique to a multiaperture source because of the difficulty of delivering O_2 to all parts of the ionizer surface and the gas load which might be involved.

All things considered, there appears to be only one type of ion, Cs^+, for which high intensity extended area sources are feasible, and for cesium the clear ionizer choice is tungsten.

9.10 A Critical Comparison with Plasma Sources

Whereas a rear-fed porous tungsten ionizer as a source of Cs^+ ions is thoroughly proven as an ion source of virtually unlimited current density, very low power requirements and mass utilization efficiencies of 99% or better, it would appear to be almost an ideal source for applications, say, to sputtering or propulsion where Cs^+ ions are acceptable. In view of this it seems necessary to emphasize that, even where Cs^+ ions are desirable or essential, a cesium plasma source would usually be preferable as a high current, extended area, ion source. This relates to the fact that the technology involved in constructing porous ionizer sources is very difficult, whereas plasma sources of Cs^+ ions are very simple to construct. Furthermore, the low ionization potential and high ionization cross section of the cesium atom yields plasma sources for which the mass utilization efficiency is commonly 90–95%, less than 99% but more than adequate for most purposes including space propulsion. In addition, with the utilization of cesiated surfaces as electron emitters the sources have lifetimes limited only by electrode erosion. Such a lifetime can be tens of thousands of hours. It would be an unusual application for which surface ionization sources would be preferable. I refer specifically to multiaperture sources. For low current single button sources surface ionization will often be preferable to plasma sources for a large number of ion species.

Chapter 10

Negative Ion Sources

10.1   The Need for Intense H⁻ and D⁻ Ion Beams

Negative ions of many elements are used as a means of obtaining highly charged states of the atoms. For example, an O^- ion beam accelerated to 10 MeV and passed through a thin foil will emerge as a beam containing ions such as O^{6+}, O^{7+}, and O^{8+}, which are very useful for increasing the effective voltage of a tandem van der Graaf accelerator. Another important use of negative ions occurs in accelerators in which ions circulate in magnetic fields. With negative ions the extraction problem is easily solved. After acceleration the beam is passed through a foil where the negative ion becomes a positive ion whose curvature is thereby reversed.

Our concern here is restricted to the generation of high current, that is beams of several amperes, of H^-, which serve quite a different purpose, that is, as a step in the production of intense neutral beams for heating magnetically contained fusion plasmas.*

Specifically, the magnetic fields which prevent the escape of plasma in a fusion reactor preclude the injection of charged particles. There have been proposals for the injection of very energetic heavy ions, whose radius of curvature in the containment fields would be sufficiently large to permit injection, but these proposals are no longer among the favored schemes for plasma heating. For light atoms, such as hydrogen or—better—deuterium since it is one of the reacting species in the favored D-T reaction, it is necessary to inject these as neutrals. If these neutrals

---

*I am aware that neutral beams are also under development for the SDI program. Because that program represents an escalation of the arms race and increases the danger of nuclear war I consider it an unworthy pursuit and will give it no further mention. Please see my supplementary preface, ''The Menace of SDI.''

are injected with sufficient energy, they can be the agents for heating the contained plasma from the ohmic heating limit near $T = 1$ keV to the approximately 10 keV temperature required to produce a sustained fusion reaction. To obtain the necessary penetration of the contained plasma before becoming ionized and thereby trapped, and for other reasons as well, deuterium energies of 400 keV or higher may be required.

Energetic neutral beams can be made from either positive ion beams or negative ion beams. Fast neutrals can be created from fast positive ions by the transfer of electrons from neutral gas molecules to the fast ions. At energies below about 20 keV for H^+ atoms the effective thickness of the charge transfer cell can always be made great enough to effect a virtually 100% conversion. For higher energies the decrease in charge transfer cross section cannot be overcome by an increase in thickness to achieve the nearly 100% conversion to neutrals because of a competition with the reionization cross section, which rises with energy. Figure 10.1 shows the limiting fraction of neutral H as a function of ion energy, plotted from data in Table A6.2 in Barnett et al. (1977). The limiting fraction for deuterium atoms should be very nearly the same as for hydrogen atoms of the same velocity and the deuterium curve displayed in Fig. 10.1 is obtained from the hydrogen curve by a displacement to the right by a factor of 2 in energy. For H^+ beams at 100 keV and D^+ beams at 200 keV the conversion to neutrals is falling so rapidly with energy that it is not feasible to use positive ion beams for the generation of neutrals at much higher energies.

The generation of neutral beams by the stripping of the outermost electron from a negative ion is not subject to any high energy limitation. The competitive process in this case is also the ionization of the neutral, but the ratio of the cross sections for the reactions $H^- \rightarrow H$ and $H \rightarrow H^+$ is independent of energy, at high energy. A neutral gas cell of the correct thickness is capable, according to Fink and Hamilton (1978), of yielding 60–65% neutrals from an incident H^- beam, independent

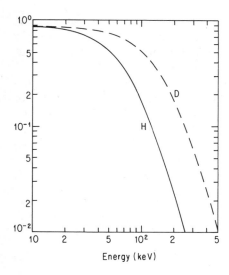

Figure 10.1. Limiting neutral fractions produced by the passage of H and D beams through H_2.

of energy. Because long range interactions are emphasized in coulomb collisions the ratio of the stripping cross section to the ionization cross section should be better in a plasma. Dimov and Roslyakov (1975) experimentally observed conversion efficiencies for H^- to H of approximately 80% in lithium and magnesium plasmas. The possibility of stripping by passing the H^- beam through an intense laser field of the correct frequency also exists, and for this process there would be no upper limit. However, the required optical intensities are so great that the process probably would have applicability only where the D^- beam has a small cross section in at least one direction.

Magnetically contained plasmas can be heated to fusion temperatures by RF as well but there are substantial difficulties. The development of a very intense, highly directional beam of D^- ions would solve an important problem and many laboratories are pursuing this problem. There are several basically different processes involved in H^- (or D^-) production and we shall consider each of them separately. The increase in H^- currents from the milliampere to the ampere range began with experiments on double charge transfer at Berkeley and Livermore. When the experiments by Bel'chenko and colleagues at Novosibirsk showed dramatic efficiencies of H^+ to H^- conversion on cesiated surfaces, the emphasis turned to surface plasma sources. Through this period Bacal and collaborators at Palaiseau reported unexpectedly high H^- density in hydrogen plasmas. This work led the way to the present emphasis on volume production H^- sources. Notwithstanding the present emphasis, all three approaches are still actively pursued and are all described in the sections that follow.

The physics of negative ions, as it was known at the time, was covered in a volume which, unfortunately, received its last revision by Massey (1976) just as the explosion in negative ion source development was beginning, together with the consequent intensive study of fundamental processes leading to negative ion formation and destruction.

10.2 Double Charge Transfer

In one approach, illustrated in Fig. 10.2, the starting point is a positive ion beam. When this beam passes through a cell containing the vapor of atoms of a sufficiently low ionization potential, a fraction will emerge as negative ions. In an alkali metal vapor such as cesium this would involve the two charge transfers

$$H^+ + Cs \rightarrow H + Cs^+$$

and

$$H + Cs \rightarrow H^- + Cs^+.$$

For an alkaline earth vapor such as strontium the H^- formation may occur in a 2-electron capture by a proton in a single collision given by

$$H^+ + Sr \rightarrow H^- + Sr^{2+}.$$

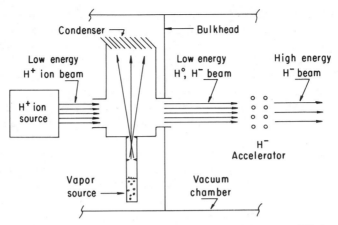

Figure 10.2 Schematic diagram of a double charge transfer source of H⁻ ions.

The physics of these double charge transfers is discussed in some detail by Olson (1976).

These reactions cannot proceed to 100% conversion simply by making the vapor cells sufficiently thick. The H⁻ ion can also lose its outer electron by charge transfer or stripping so that there is an equilibrium fraction F^∞ describing the maximum fraction of the incident beam which is converted to negative ions. Some values of F^∞ for hydrogen or deuterium ions in three alkali metal vapors from Schlachter et al. (1980) are shown in Fig. 10.3, and for the alkaline earth vapors from McFarland et al. (1982) and Schlachter et al. (1983) in Fig. 10.4.

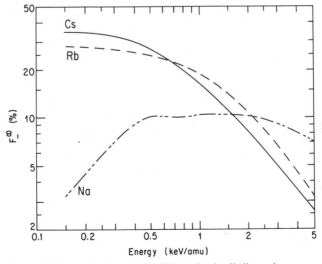

Figure 10.3. Equilibrium yields of H⁻ or D⁻ in alkali metal vapors.

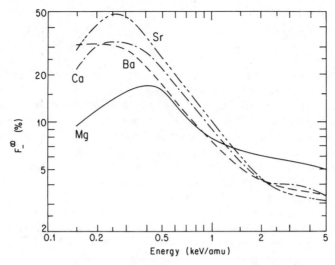

Figure 10.4. Equilibrium yields of H⁻ or D⁻ in alkaline earth vapors.

A competing process, the stripping of an electron from the neutral atom to form an H^+ ion, is unimportant in the energy range plotted in Figs. 10.3 and 10.4. Where this process becomes significant there is an optimum thickness, and the fraction of incident positive ions which become converted to negative ions for this thickness is the parameter η_-^{opt}. For the energy range less than 5 keV/amu plotted in Figs. 10.3 and 10.4, the parameters F^∞ and η_-^{opt} are virtually identical.

For a jet of adequate thickness H^+ would be virtually absent from the low energy beam leaving the jet. It is necessary in that case to be concerned about space charge neutralization of the low energy H^- (or D^-) beam in the space between the vapor jet and the high voltage accelerating electrodes. Although this has been the subject of much concern, it should present no problem because the positive ions produced in the vapor jet, Cs^+, for example, would diffuse into the negative space charge beam where they would be trapped in whatever numbers are required to space charge neutralize the H^- beam. Similarly, the high energy H^- beam will probably receive adequate neutralization from positive ions produced in the stripping cell, providing the beam energy per ion is 50 keV/amu or greater. Anderson and Hooper (1977) calculate adequate neutralizing ions produced in the beams themselves, although this neutralization process requires an adequate ambient gas density. They do not take into account the neutralization to be expected from the positive ions that drift into the beam from the stripping cell.

An important consideration in the acceleration of the negative ion beam is whether the beam plasma between the charge transfer cell (the vapor jet in Fig. 10.2) and the negative ion accelerator contains many electrons. The flow of electrons from the H^+ ion beam plasma through the vapor jet will be suppressed because the H^+ beam plasma potential and the potential in the jet where slow charge transfer positive ions are being created will be positive relative to the potential in the H^- beam leaving the vapor jet. Anderson and Hooper (1977) calculate that even if there is a large electron density in the charge changing cell very few electrons will reach the negative ion accelerator.

Before proceeding to describe various attempts at producing H^- or D^- beams by double charge transfer it is worth noting a uniquely attractive feature of this method; the pumping requirements for the negative ion and neutral beam portion of the system can be substantially reduced, as pointed out by Anderson (1975), by using the vapor jet, whose primary function is to produce the desired charge transfer, as a barrier to the flow of neutral gas from the positive ion source.

As shown by Hooper and Willman (1977) the cross sections for charge transfer are sufficiently large compared to cross sections for scattering into large angles that the negative ion beam to the right of the jet in Fig. 10.2 can be expected to have only slightly more angular spread than the positive ion beam, for a jet thickness adequate to achieve a negative ion fraction close to F_-^∞. A neutral gas molecule, on the other hand, which might be traveling at the order of 5×10^{-3} times the speed of even a 500 volt H^+ or D^+ ion, would spend so long a time in the jet that it would be carried along by the vapor. The target thickness required to carry the positive ion density to near zero and the D^- density to its equilibrium value was shown by Schlachter et al. (1980) to be about 10^{15} cm^{-2} for 1 keV and 2.5 keV D^+ ions through a Cs jet and by McFarland et al. (1982) to be about 5×10^{15} cm^{-2} for 800 eV D^+ ions through Sr and 3 keV D^+ ions through Ba. Such vapor streams would act like the jets of diffusion pumps, preventing gas from the positive ion source from following the beam ions. A chamber bulkhead to the right of the vapor jet, as shown in Fig. 10.2, would effectively keep gas from the H^+ (or D^-) source out of the negative ion portion of the system.

Among the various charge exchange media shown in Figs. 10.3 and 10.4, strontium stands out for its extremely high yield of negative ions. However, there are difficulties working with a strontium jet. At its melting point, 1043 K, its vapor pressure is high, about 1.6 torr, so that recycling by a liquid flow from the condenser back to the vaporizer is not feasible. The requisite target thickness (5×10^{15} atoms/cm^2) and the requisite evaporation temperature (~ 800 K) can readily be translated into a requirement of an evaporation rate of about 10^3 cm^3/hour for a beam 10 cm in diameter. It would be difficult to store enough for a long run. Similar considerations apply to all of the alkaline earth materials.

The alkali metals are much easier to work with. Cesium, for example, melts at 30°C where its vapor pressure is 2.5×10^{-6} torr and would produce an adequate flow of vapor at temperatures of 150–200°C. Sodium melts at about 100°C where its vapor pressure is only 10^{-7} torr and an oven temperature of 300–350°C would be adequate. These materials are compatible with stainless steel and are subject to electromagnetic pumping for convenient recycling.

Of the two alkali metals cesium and sodium, cesium shows a much higher peak value of F_-^∞, nearly 35% at 200 eV/amu, falling off slowly out to 500 eV/amu. Although sodium has a peak value of F_-^∞ which is much lower, the value holds up to much higher voltages. For example, at 5 keV/amu, $F_-^\infty = 0.07$, whereas the value for cesium at this energy is only one-tenth as large. This is an important consideration because of the relative ease of making 10 keV D^+ beams compared to 0.4 to 1 keV D^+ beams.

The earliest experiments on double charge exchange generation of D^- appear to be those of Osher et al. (1972), who converted a 200 mA, 1.5 keV D^+ ion beam to 42 mA of D^- ions. The work with cesium, however, met with the difficulty of obtaining a high current density, low energy D^+ ion beam with low angular spread using conventional electrodes. It is possible to obtain large current densities at low ion energies by having an accel gap voltage much larger than the net voltage between the source and ion beam, but such arrangements invariably lead to undesirably large angular spreads. Because of this difficulty Semashko et al. (1977) and Hooper et al. (1981) turned to experiments with sodium as the charge changing medium.

The D^+ ion source of Hooper et al. was that shown in Fig. 8.15 with an accel electrode consisting of 105 slots, each 2 mm $\times$ 7 cm. The angular divergence of the ion beam was about $0.75°$ in the direction parallel to the slots and about $2.75°$ in the direction perpendicular to the slots. As expected, the angular spread changed very little in the charge transfer cell.

The charge transfer cell contained a well collimated vapor jet, whose profile is shown in Fig. 10.5, produced by a supersonic flow nozzle. The D^- current density achieved by Hooper et al. is shown in Fig. 10.6 as a function of the effective thickness of the sodium cell. The efficiency with which the directed D^+ ions were converted to directed D^- ions is displayed in Fig. 10.7. The summation of their achievement is a total D^- current of 2.2 A at a peak current density of 15 mA/cm^2 in the range of 7–13 keV, with an angular divergence of $0.75° \times 3°$.

It should be noted that a primary reason for requiring the low energy D^+ beam

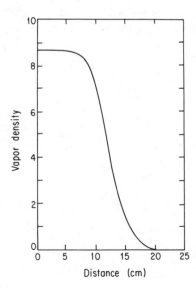

Figure 10.5. Sodium vapor density profile of Hooper et al. (1981) for $nL = 2 \times 10^{15}$ cm^{-2}. The vapor density is in arbitrary units and distance is measured from the center line of the jet.

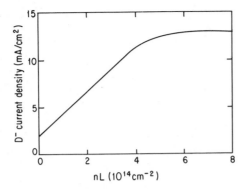

Figure 10.6. D^- current density as a function of effective target thickness obtained by Hooper et al. (1981) for 10.5 keV D^+ ions.

to have a small angular spread has less to do with the transverse energy than it does with the size of the beam at the vapor jet. For example, a 0.1 radian angular spread in a 500 eV beam would have transverse energies up to 5 eV. When accelerated to 300 keV this would yield an angular spread of 4×10^{-3} radians or $0.23°$.

Semashko et al. (1977), working with a 10 keV hydrogen ion beam and a sodium target, converted a beam of positive ions and neutrals with an equivalent current of 8 A, measured calorimetrically, to 1.4 A of H^- ions. This extraordinary efficiency is attributed to the fact that their positive ion source produced a high percentage of molecular ions, 24% H_2^+ and 22% H_3^+, with only 54% of the beam as protons. When a 10 keV H_2^+ ion receives a charge transfer electron it will be in an excited state which can fly apart making two 5 keV H atoms, or for some excited states a 5 keV proton and a 5 keV H^- ion. Similarly, the H_3^+ ion can lead to three 3.3 keV H atoms and ions. De Bruijn et al. (1984) measured large cross sections for the dissociative charge exchange of H_2^+ in Cs and Na, even though their experiments excluded reactions in which the excited neutral molecule disintegrated directly into H^+ and H^-.

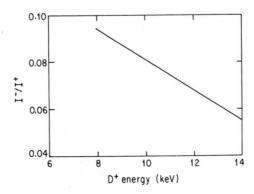

Figure 10.7. Conversion efficiency of D^+ to D^- in sodium vapor as a function of ion energy.

In later experiments Semashko et al. (1986) achieved H^- currents as high as 5.5 A and accelerated these ions to a energy of 80 keV with a beam cross section of 5 cm $\times$ 40 cm. The pulse lengths used were 10–25 ms and were limited to no more than 0.3 second by their available pumping speed.

In these experiments Semashko et al. passed a 9 keV H^+ ion beam through a supersonic Na vapor jet. The 9 keV energy was chosen as the optimum based on an available positive ion current rising approximately as the $3/2$ power of energy and the decreasing charge transfer cross section. The conversion efficiency $J-/J+$ was about 10%, down from earlier experiments presumably because of a smaller fraction of molecular ions in the source. For the positive ion source which produced the highest H^- currents, the H^- beam out of the target contained about 48% 9 keV ions, those created from H^+ ions, 26% 4.5 keV ions from H_2^+ ions, and 26% 3 keV H^- ions originating from H_3^+ ions, at H^- currents up to 3 A of H^-. At 5 A these percentages had changed to 42, 40, and 18%, respectively. It can be estimated that 50–75% of the H^- ions are produced from molecular ions.

The sodium vapor jet had a linear density of 0.5×10^{15} to 2.0×10^{15} cm^{-2} and proved, as expected, to be an effective barrier to gas flow. With a pressure of 10^{-4} to 10^{-3} torr on the positive ion source side of the jet the pressure was only 10^{-6} to 10^{-5} torr on the negative ion side. The plasma density in the vapor jet under high current conditions would reach 6×10^{12} cm^{-3}. The charge transfer between the Na^+ ions and streaming neutrals caused a deceleration of the jet and this in turn caused an enhanced leakage of sodium vapor. This produced a sodium coating on the H^- accelerating electrodes, which created voltage breakdown problems. These were overcome by maintaining the electrodes at a temperature of 100–120°C.

The potential in the sodium jet plasma was +4 volts and the electron temperature only 0.7 eV. In the H^- beam the potential was positive but less than 1 volt so that the potential difference tended to prevent electrons from flowing into the H^- beam. Nevertheless, it was necessary to take further steps to prevent the acceleration of a large electron current along with the negative ions. A crossed magnetic field of 85 gauss plus a negative potential on the first electrode facing the charge transfer cell was able to accomplish this. For H^- current less than 2 A, 10–20 volts was adequate for cutoff of the electrons. At 3.5 A the requisite voltage was 400 volts and the paper of Semashko et al. is not clear on what was done for still higher currents.

The H^- accelerator was a single large aperture. Presumably a gridded structure would have made feasible a low electron cutoff voltage out to much larger currents. A 5.5 A 80 kV H^- ion beams, with a cross sectional length to width ratio of 8, has a poissance Π, defined by Eq. (2.116), of 0.56, so that space charge effects can be moderately important in the acceleration and propagation of this beam. Since any spreading comparable to the $\Pi = 0.6$ curve of Fig. 2.15 would have been readily observed, it can be assumed that the 80 keV H^- beam was space charge neutralized by positive ions produced by interaction of some of the fast H^- ions with the ambient gas.

It is important to have the beam size as small as possible at the vapor jet. The

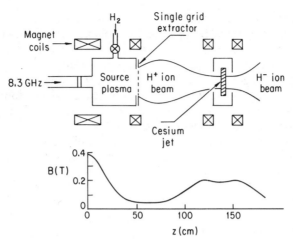

Figure 10.8. Schematic of the double charge transfer experiment of Delaunay et al. (1983).

larger the cross section of the jet, the harder it becomes to make the jet narrow and therefore the further it must be removed from the source lest the charge changing vapor, the cesium or sodium, for example, interfere with the source operation.

Geller et al. (1980) and Delaunay et al. (1983), whose experiment is shown schematically in Fig. 10.8, attacked the problem of obtaining a small diameter beam at their cesium vapor jet by the compression of an axial magnetic field in which the source operates and on which the beam is trapped to a diameter compatible with a supersonic cesium jet developed by Bacal et al. (1982).

This source was an RF source driven by about 8 kW of power at 8.3 GHz which strongly coupled to the plasma through electron cyclotron resonance at 0.3 tesla at the rear of the source. The field fell to about 500 gauss at the front of the source where the ions were extracted by a single grid extraction system as described in Sec. 5.10F. Their extraction area was 14 cm × 14 cm and was covered with a 50% transparent array of 0.5 mm diameter holes. At the low plasma density at which they operated the plasma delivered 20 mA/cm^2 to the grid for an extracted current density of 10 mA/cm^2, of which 85% was H$^+$. The plasma was held at 300–500 volts relative to the grid. The angular spread is readily estimated from Eq. (5.18). Taking the beam to be 10% H$_2^+$ and 5% H$_3^+$ we obtain an effective value of $\chi = 5.1 \times 10^{-8}$ A/V$^{3/2}$ from which we obtain an estimated angular spread of 4.2° at 300 volts and 2.8° at 500 volts. The compression of the area by a factor of 4 as the beam is transported from the extraction grid to the jet by a magnetic field increasing from 0.05 to 0.2 tesla will double this angular spread. Because the conversion of H$^+$ to H$^-$ is a two stage process, with the neutrals and charged particles following different paths, the emittance of the beam will be increased in the jet, independent of scattering effects. For the thin cesium jet developed by Labaune (1978) and Bacal et al. (1982) this effect might be small.

To eliminate electrons from the H$^-$ beam prior to acceleration Delaunay and colleagues propose that the magnetic field be rapidly reduced to zero before acceleration to high voltage. This would carry all electrons out of the beam and permit the H⁻ ions to continue their trajectories but not without a sudden lateral kick described by the Busch theorem. The azimuthally directed velocity given an ion in this process is

$$v_\phi = \frac{1}{2} \frac{e}{M} rB \qquad (10.1)$$

and the equivalent voltage obtained by equating

$$\tfrac{1}{2} M v_\phi^2 = e V_\phi \qquad (10.2)$$

leads to

$$V_\phi = \frac{1}{8} \frac{q}{M} r^2 B^2. \qquad (10.3)$$

For an ion 8 cm from the axis in a field $B = 0.05$ tesla this would lead to $V_\phi = 192$ volts for H⁻ and 96 volts for D⁻. After acceleration to, say, 300 kV this would yield an angular divergence of only 1° due to this effect for a D⁻ beam. The complications of this approach are formidable and, judging from the absence of any reports at the 1986 Conferences on Negative Ions at Palaiseau and Brookhaven, the experiment appears to have been abandoned.

A technique which has not been adequately pursued for double charge transfer generation of D⁻ depends on the use of a fine mesh extraction of D⁺ ions. For example, at 500 volts a current density of 0.23 A/cm² would bring the grid to the safe temperature for tungsten of 2200 K, from considerations discussed in Sec. 5.10. If the grid is 70% transparent and the beam is 85% D⁺, we would then have a D⁺ ion beam of 137 mA/cm². An 85% D⁺ beam, with 10% D_2^+ and 5% D_3^+, would have an effective value of $\chi = 3.4 \times 10^8$ A/V³/² and Eq. (5.18) would yield an angular spread of 4.27°. If the beam originated at a 10 cm diameter extraction area and the grid were contoured, as illustrated in Fig. 5.31, such a beam could be kept to no more than 10 cm diameter for a distance of 50 cm where it could intersect a cesium jet. The 11 A D⁺ beam entering the jet would be expected to emerge as a 3.6 A D⁻ beam ready for acceleration to high voltage. The disadvantage of this system would be the fragility of the grid, as discussed in Sec. 5.10F.

10.3   Surface Production of H⁻ or D⁻

A.   First Observation of High Current Density Surface Ionization

A plasma contains not only electrons and positive ions, but negative ions as well, whenever the atomic or molecular species has stable negative ion states. The extraction of negative ions from a plasma volume will be discussed in greater detail in Sec. 10.4, but it is clear that the fields that extract and accelerate negative ions will also extract and accelerate electrons, whose flux density in the plasma is vastly greater than that of negative ions, unless the electrons are restrained by magnetic fields. Plasmas used for negative ion generation, then, need to operate in magnetic fields or, at least, have magnetic fields across the extraction apertures which strongly inhibit electron flow, while passing negative ions with an acceptable amount of deflection.

A convenient geometry for creating a plasma in a magnetic field is that shown in Fig. 10.9 which Bel'chenko et al. (1973, two papers, and 1974) term the planotron. One can view this configuration as a flattened out magnetron configuration. The cathode potential end plates also provide an oscillating electron character to the region in which the electrons parade around the central cathode. As in other cold cathode discharges the cathode emission results from secondary emission under ion bombardment, but whereas most cold cathode discharges run at fairly low plasma densities, this configuration is capable of creating dense plasmas in a magnetic field of 0.1 tesla hydrogen densities of 10^{16} molecules/cm^3 (approximately 0.3 torr) and voltages of 500–600 volts. Pulse lengths were kept to 10^{-3} second to avoid significant cathode heating. In pure hydrogen, H$^-$ yields up to 0.75 A/cm^2 were obtained from a 0.4 mm $\times$ 1 cm slit. When cesium vapor was added the requisite H$_2$ density dropped to 3×10^{15} cm^{-3}, the arc voltage to 100–150 volts, and the extracted H$^-$ density rose to 3.7 A/cm^2. The source pressure with cesium is given in a later paper by Dudnikov (1980) as approximately 5 Pa $\approx$ 0.04 torr, which would correspond to a density of about 1.2×10^{15} cm^{-3}.

The energy distribution of the H$^-$ ions showed two peaks, a sharp peak corresponding to ions formed at plasma potential and a broad peak of ions more energetic than this by amounts of the order of the arc voltage. The explanation lay in the generation of H$^-$ ions at the cesiated molybdenum cathode surface. The H$^+$, H$_2^+$, H$_3^+$, and Cs$^+$ ions from the discharge get acclerated across the cathode sheath to the cathode surface. They may reflect with anything from zero to their full energy, the molecular ions may decompose into atoms, and they may dislodge adsorbed hydrogen from the surface. Particles cannot leave the surface as positive ions but a fraction of the atomic hydrogen leaving the surface does so as H$^-$ ions. Since these leave the cathode with energies between zero and the sheath voltage, which is approximately the arc voltage, after acceleration back across the sheath

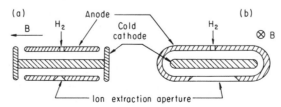

Figure 10.9. Simplified sectional views of the planotron (a) along the magnetic field B and (b) normal to B.

they would have a total energy, in electron volts, which would range from the arc voltage to twice that value. As these H⁻ ions pass through the plasma some undergo a resonant charge transfer to atomic hydrogen, thereby creating the slow H⁻ ions which formed a significant component of the H⁻ spectrum by Bel'chenko and colleagues.

B. Theory and Basic Experiments

The process by which an H atom leaving a cesiated surface becomes an H⁻ ion presumably is similar to the surface ionization process by which a Cs^+ ion is formed at a tungsten surface. In this case the Saha–Langmuir equation (9.5) would become

$$\nu_n/\nu_a = (g_n/g_z) \exp\left[e(I_a - \phi)/kT\right], \qquad (10.4)$$

where ν_n is the flux of negative ions from the surface and I_a is the electron affinity of the atom. The application of this equation to the formation of negative ions when plasma ions strike a surface is questionable since, unlike the case of cesium evaporating from a tungsten surface, the hydrogen atoms leaving the surface are not in thermal equilibrium with the surface. For the moment ignore this important difference and consider the implications of Eq. (10.4). The most obvious aspect of this equation is that the production of a large fraction of H⁻ among the H atoms leaving the surface would require that the electron affinity of H be comparable to, or greater than, the work function of the surface.

Graham (1980) gives a survey of work functions of many metals, polycrystalline and single crystal faces, when coated with absorbed layers of the alkali metals. Values as low as 1.45 were obtained for cesium on polycrystalline rhenium and on the (1 1 0) face of tungsten. The lowest value reported for cesium on molybdenum, the combination which appears to have been most successful, is 1.54 volts.

The relevance of these data is uncertain because the surfaces will contain absorbed hydrogen as well as the alkali metal atoms. Papageorgopoulos and Chen (1973) measured the work function produced by the delivery of cesium to a hydrogen coated tungsten (1 0 0) crystal plane and obtained a value of $\phi_{min} = 1.42$ volts. This value is not expected to vary much from one crystal plane to another, or even to be much different for molybdenum.

The electron affinity of hydrogen is only 0.75 volt and the difference ($\phi_{min} - I_a$), using the Papageorgopoulis and Chen value for ϕ_{min}, is 0.67 volt. At a temperature such tht $kT/e = 0.05$ volt, at which the Bel'chenko cathodes ran, Eq. (10.4) would yield a fraction H⁻/H $\approx 8 \times 10^{-7}$, taking the ratio of the weighting factors as $1/2$. Palmer (1983) observed yields under thermal equilibrium conditions of about 10^{-3}, approximately 10^3 times larger, which he attributes to strong chemical reactions between the surface, the cesium, and the hydrogen. Even this H⁻ yield is much too low to explain the spectacular yields obtained with the H⁻ source of Bel'chenko et al.

There are substantial errors in the use of Eq. (10.4). Consider first the neglect of the sheath field at the cathode surface. This field can be estimated readily from considerations covered in Sec. 3.8 to be about 10^5 V/cm. This can lower the work function, as illustrated in Fig. 7.6, through the so-called Schottky effect. As mentioned in Sec. 7.3, composite surfaces show a very large Schottky effect so that the effective work function will be substantially lowered by the sheath field. A lowering of 0.25 volt would be consistent with the magnitude of electron current increases obtained from composite surfaces in electric fields.

Another effect that can be produced by a surface field is illustrated in Fig. 10.10. At distances from the surface where tunneling is still possible an electron on an H^- ion will be in a lower state than on the surface. The effect is almost certainly negligible. If tunneling from the Fermi level in the metal occurs with significant probability out as far as 10 angstroms, this would produce an effective lowering of the energy level of the H^- extra electron by only 10^{-2} volt in a field of 10^5 V/cm.

The use of Eq. (10.4) in any manner is probably erroneous but if it is to be used, then a temperature other than the temperature of the emitter must be used. In explaining sputtering and secondary electron emission due to an ion bombardment a *thermal spike model* is often invoked. Specifically, it is assumed that the energetic ion creates a microregion of very high temperature and that the secondary electrons are emitted thermionically, and the sputtered atoms are thermally evaporated from the hot microregion. In experiments by Yu (1977), in which he measured H^- currents from a Mo–Cs–H surface bombarded by Ne^+ ions, he did find an H^- current which varied as $\exp(\Delta\phi/E_0)$, where $\Delta\phi$ is the lowering of the work function and E_0 was a measured parameter. For Ne^+ energies between 150 and 2000 eV he found E_0 very close to 0.5 volt, corresponding to a temperature of 6000 K. Schneider et al. (1977) measured the D^- yield from solid alkali metal surfaces bombarded with D_2^+ and D_3^+ ions. Again they found that the yield decreased as the work function increased but to match the yields would require values of kT/e in the range of 0.4–0.8 volt for bombarding ion energies of 100–400 eV/nucleon. Huffman and Oettinger (1980) also found satisfactory agreement with the thermal spike model. Their effective temperature for a cesium surface bombarded with 100–400 eV hydrogen ions was 9547 K, based on Eq. (10.4) and a work function of 1.90 eV. A lower assumed work function would lower the effective temperature.

If, in Eq. (10.4), we use $kT/e = 0.5$ volt, with $(\phi - I) = 0.67$, we obtain $H^-/H = 0.13$. If the anomalous Schottky effect due to the sheath field were to lower the work function by 0.25 volt, the ratio would become a substantial 0.22.

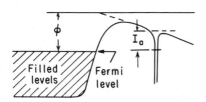

Figure 10.10. Potential energy diagram for an atom escaping from a surface in an electric field. I_A is the electron affinity of the atom and ϕ is the work function of the surface.

If reasonable agreement with experiment were the criterion for acceptance of a model, the use of the Saha–Langmuir equation together with the Schottky effect and the thermal spike model would appear to be an adequate description of the phenomenon. In fact many highly detailed analyses of the phenomenon have been undertaken, for example, by Bel'chenko et al. (1977), Hiskes et al. (1976), and Hiskes and Karo (1977). A paper by Lang (1983) derives the fractional ionization defined by

$$\eta = \nu_{H^-}/(\nu_{H^-} + \nu_{H^0}).\qquad(10.5)$$

For high particle escape velocities he finds that

$$\eta = \exp\left[(I_a - \phi)/cv_\perp\right],\qquad(10.6)$$

where c is a constant and $v_\perp$ is the normal component of the particle velocity as it leaves the surface.

I shall not attempt to cover the theory leading to Eq. (10.6), but a few words to make this equation, which is so different from Eq. (10.4), seem reasonable, are in order. When a hydrogen atom is close to the plane at which the potential energy in Fig. 10.10 is equal to the Fermi level, the level representing the extra electron on the negative ion will be below the Fermi level and greatly broadened by the interaction with the conduction electron levels in the metal. There is therefore a high probability that the state will be filled. As the atom moves away from the surface the level will rise above the Fermi level and, to the extent that tunneling permits, the probability of the electron occupying the state decreases. Equation (10.4) inherently assumes that the atom moves so slowly that equilibration with the conduction band takes place out to such a distance tht the atom/ion is free of the surface. If the atom moves away from the surface rapidly, then the probable occupancy of the negative ion state will be higher, corresponding to a point closer in.

Although it is likely that Eq. (10.6) gives an improved model, it does not appear to be the last word. Granneman et al. (1983), for example, found a strong dependence of the probability of a particle escape as a negative ion to be strongly dependent on the component of velocity parallel to the surface.

A substantial amount of experimental work has been undertaken on the surface production of H⁻ ions but the field is still in a developing stage and I have not found it feasible to organize the results in as orderly a fashion as for a field in a more mature state. Nevertheless, it seems important to present the results of experiments that have been carried out. Leung and Ehlers (1983) studied the emission of H⁻ ions from various metals coated with cesium. From the energies of the emitted ions they were able to conclude that more ions were produced by bombardment activated desorption (sputtering) than from backscattered or reflected particles. In fact, of the numerous converter surfaces they studied only copper and stainless steel showed significant backscattered H⁻ ions. They found that molybdenum was superior in its H^- production to tantalum, copper, stainless steel, and niobium, but that titanium and vanadium were slightly superior to molybdenum.

Geerlings et al. (1986) found that the negative ion yield per nucleon from a cesiated tungsten (1 1 0) monocrystalline surface was the same for H^+, H_2^+, and H_3^+. They found that there were no H^+ ions among the reflected particles. The efficiency of H^- generation could be fitted by

$$\eta = \exp\left(-1/\beta v_\perp\right) \tag{10.7}$$

where β is a constant, in accord with the theory of Lang (1983) and with theoretical work of Gauyacq and Geerlings (1986). Values of η as high as 0.67 were found for 100 eV protons striking a 0.6 monolayer cesiated tungsten (1 1 0) surface at an angle of incidence and reflection of 70°. Surprisingly enough, in view of the finding of Papageorgopoulos and Chen (1973), exposure of the cesiated surface to hydrogen greatly reduced the H^- yield.

Alessi et al. (1984) argue that the rate at which cesium atoms are deposited in an intense plasma and the rate at which Cs^+ ions of energy 100 eV or greater sputter the surface lead to a surface coverage below the optimum desired for minimum work function. To overcome the difficulty they carried out experiments in which liquid cesium was forced through a porous molybdenum converter to obtain the desired coverage. In a small steady state hollow cathode discharge source, of a type described in Sec. 8.3E, the H^- yields were five times higher with the rear-fed porous molybdenum converter than with a solid molybdenum converter relying on cesium coverage by vapor deposition from the plasma.

Much lower work function surfaces than the Mo–H–Cs surface appear possible. Huffman and Oettinger (1980), for example, suggest trials with a surface prepared by evaporative coating of a surface from a Cs_2CO_3 source. At 875 K, CO_2 is evolved and the less volatile Cs_2O condenses on the desired surface. The photo-electrically evaluated work function of this surface is approximately 1.1 volt. The surface can be laid down in a thick layer so that it is likely to be able to stand up to the bombardment of a cathode surface in a surface plasma source for long periods of time. I am not aware that this surface has ever been tried in H^- surface plasma sources, and probably with good reason. It would probably require very high vacuum conditions and preparation of the surface *in situ*.

In this discussion we considered the yield of H^- ions from surfaces bombarded with H^+, H_2^+, and H_3^+ ions. When cesium is introduced the plasma will also contain Cs^+ ions and Greer and Seidl (1983) measured the H^- yield from a Mo–H–Cs surface bombarded with Cs^+ ions. The yields were very small for Cs^+ ion energies <200 eV but rose rapidly with energy to a peak value of 0.55 at an energy of 750 eV.

C. Magnetron-Type Surface Plasma Sources

Returning to our discussion of surface plasma sources, Fig. 10.11 illustrates a feature of the planotron source not shown in Fig. 10.9. A plenum was placed

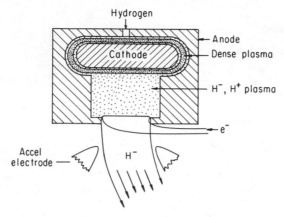

Figure 10.11. Planotron with plenum between the cathode and the extraction slit.

between the extraction slit and the intense racetrack shaped plasma region surrounding the cathode. Since electrons move across the field with difficulty compared to H⁻ ions, the plasma in this region is particularly rich in H⁻ ions, reducing the drain current of electrons at the extraction slit. In addition, increasing the length over which charge transfer occurs increases the current density in the monoenergetic peak at the expense of the H⁻ ions with a broad energy spectrum.

Bel'chenko and Dudnikov (1979) built another source they termed the semiplanotron in which the plasma for converting fast H⁻ ions to slow H⁻ ions was eliminated in order to maximize the extracted current. The cathode in this source was grooved to focus the surface produced H⁻ ions through the extraction aperture. This source evolved at Novosibirsk to the pulsed multiampere source of Bel'chenko and Dimov (1983), but it will be convenient first to describe developments at Brookhaven achieved by Prelec (1980) and Alessi and Sluyters (1980).

Prelec and Alessi and Sluyters observed an improvement in performance produced by a focusing groove, and another improvement produced by widening the back side of the racetrack discharge gap as suggested by Wiesmann (1977). The latter improvement is due to the fact that opening the region permits operation at a lower pressure. The configuration of Alessi and Sluyters is shown in Fig. 10.12 and the gains in performance produced by the changes are illustrated in Fig. 10.13. The source was uncooled and pulse lengths were limited to 10 ms on a very low duty cycle. An important gain was a change in the gas utilization efficiency from 2 to 6%.

With D_2 instead of H_2 the currents were the same. This may seem surprising but the same arc current can be expected to yield a plasma density greater by a factor of $\sqrt{2}$ for deuterium; this would produce the same sheath current as for hydrogen. Although the velocity of the ions striking the cathode would be less for the deuterium ions, the energy per ion would be the same and, in the thermal spike model, the localized temperature produced would be similar. The energy of the

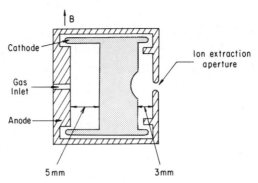

Figure 10.12. Cross section of the Brookhaven magnetron with focusing groove and enlarged rear space.

H⁻ ions observed by Alessi and Sluyters showed two peaks, one corresponding to emission at low velocity from the cathode and the other originating at low velocity in the plasma. For some reason the surface used by them did not reflect the H atoms/ions. Rather it appears that they were adsorbed and subsequently sputtered from the surface.

Another interesting experiment performed by Allesi and Sluyters was a test with the extraction slit parallel to the magnetic field rather than perpendicular to B. For reasons which are not clear this degraded the performance by a factor of 2.

Witkover (1983) reported on the use of this source with the Brookhaven Linac

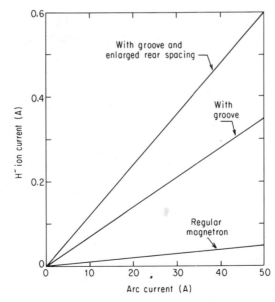

Figure 10.13. H⁻ current from a 0.6 mm × 45 mm slit as a function of arc current for three configurations of the Brookhaven Mark III magnetron.

for over 18 months with H⁻ currents of 40 mA delivered in 0.6 ms pulses operated at five pulses per second.

The magnetron type of surface plasma source of negative ions was used at several other laboratories as well. At Los Alamos, Allison et al. (1980) obtained an H⁻ current of 0.1 A from a 1 cm × 0.05 cm slit. Their measurements indicated a ratio of negative ions leaving the cathode to the positive ion current striking the cathode of 0.7, substantially higher than other results. The same group, as reported by Smith et al. (1980), experimented with a rotating cathode in order to increase the duty cycle. They were able to achieve a 100% duty cycle at 1 mA of H⁻ ions, but only 1% at 100 mA.

The pulsed multiampere source (PMS) previously referred to is described by Bel'chenko and Dimov (1983) and illustrated in Fig. 10.14. In this source the cathode is 3 cm in the direction of the magnetic field by 18 cm in the perpendicular direction. Instead of slits they have gone to spherical indentations in either a square or hexagonal array. In the latter case the cathode configuration forms a honeycomb array of ridges identical to the sastrugi ionizer configuration shown in Fig. 9.12.

The source was uncooled and therefore could be run only in a pulsed mode. The beam pulse lengths were 0.2–0.8 ms, but during this time an H⁻ beam current of 11 A was obtained at an arc current of 700 A, for an average current density over the total beam area of 0.18 A/cm^2. At the magnetic field of 0.5–1.5 kG and the extraction voltage of 25 kV the excursion of the electrons in their trochoidal orbits was large enough that many were collected on the accel grid. The power drain to this electrode and the cathode were both about 1 kW/cm^2.

The thickness of the zone of electron oscillation at the contoured surface, determined by the overhang seen in Fig. 10.14a, was 1–2 mm, but in the ignition gap where the gas is introduced into the discharge region the cathode projections are 4–5 mm long in order to lower the density required to strike and maintain the discharge. This, plus the fact that the current from the cathode is compressed by a factor of about 20 before it enters the accel gap, produced a pulsed gas efficiency of at least 20%.

D.   Penning-Type H⁻ Ion Source

Following the spectacular results with the planotron, Dudnikov (1974) introduced cesium into a Penning geometry source (shown schematically in Fig. 10.15) and

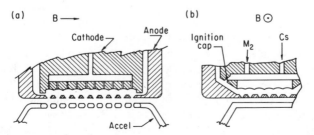

Figure 10.14. The PMS source of Bel'chenko and Dimov (1983).

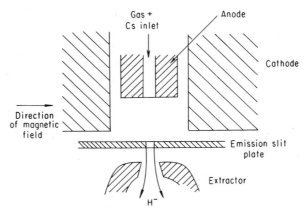

Figure 10.15. Penning-type of H^- ion source.

also obtained an improved H^- extraction. Unlike the planotron, the normal to the cathode surface is not in the direction of ion extraction. If, in fact, the H^- ions from this source are the result of surface production, then the extracted ions must be those that charge transfer with neutrals near the extraction aperture or ions emitted from the anode. Bel'chenko et al. (1977) suggest that H^- ions are formed at the anode as a result of bombardment by fast neutrals. These may be the result of ions which are reflected from the cathode with full energy, after picking up an electron, or they may be the product of a charge transfer of an electron from an H^- ion after it is accelerated across the sheath. It is to be noted, however, that the work of Jimbo et al. (1986) and Leung et al. (1986) described in Sec. 10.4D throws some doubt as to the interpretation of this source as based on the surface rather than volume generation of H^- ions. An example of the performance of this type of source is given by Allison (1977), who achieved a current of 108 mA through a 1 cm $\times$ 0.05 cm slit for 0.7 ms. His duty cycle at this current is limited by thermal effects to 0.5% and his gas efficiency was only 0.8%.

E. Modified Calutron or SITEX Source

The concept of surface plasma sources inevitably led to its application to numerous other plasma configurations, as we shall see in this section and in the two sections to follow. One source with features resembling the magnetron sources was the Oak Ridge modification, shown in Fig. 10.16, of the calutron source shown in Fig. 8.11, described by Dagenhart et al. (1980, 1981, 1983). They termed this source SITEX as an acronym for surface ionization with transverse extraction.

Aside from the admission of Cs as well as H_2 (or D_2) SITEX differs from the calutron in that the back of the chamber is not at anode potential but is an insulated plate which can be run at an arbitrary potential. Two possible advantages relative to the magnetron sources of the Novosibirsk and Brookhaven groups are evident. The first of these is the use of a thermionically emitting cathode. This should make it possible to obtain a dense plasma at lower pressure and lower voltage. The

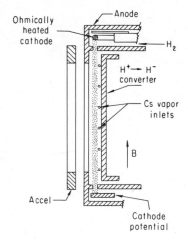

Figure 10.16. SITEX ion source.

second possible advantage relates to the independent control of the converter voltage relative to the main plasma column from which the positive ions originate. The extraction slit in this source is parallel to the magnetic field, at right angles to the direction customarily used in the magnetron sources. This was found by Alessi and Sluyters (1980) to yield only half as much negative ion current as slits perpendicular to the magnetic field.

After initial runs with a flat molybdenum converter the converter was contoured with cylindrical grooves and achieved currents of 650 mA of H⁻ ions at 18 keV for 10 second pulses with a cooled converter. The obtainable current was shown to be proportional to the extraction aperture area and the current density achieved was 130 mA/cm². The electron current drain, which was very high in initial experiments, was reduced to only 15% of the H⁻ current by increasing to 6 mm the spacing between the main discharge column and the electrode through which the ions were extracted. The electrons that did leave the discharge were paraded across the field and 90% of these were collected at a potential 0.1, the potential across the accelerating gap.

Reported performance with D⁻ was not quite as good as with H⁻. A D⁻ current of 260 mA at 100 mA/cm² in 5 second pulses was reported.

The currents quoted are the extracted currents. The currents actually reaching a collector were substantially reduced by charge transfer. The gas efficiency of the source was only 3% and the 650 mA of H⁻ ions became only 200 mA to a collector 23 cm away. It appears necessary to improve the gas efficiency, chamber pumping, or both.

F.   Hollow Cathode Discharge Sources

A different* approach to surface plasma H⁻ generation was pursued at Brookhaven by Hershcovitch and Prelec (1980, 1981) and Hershcovitch et al. (1986). Their

---

*See my comments at the end of this section pointing out the superficiality of the differences.

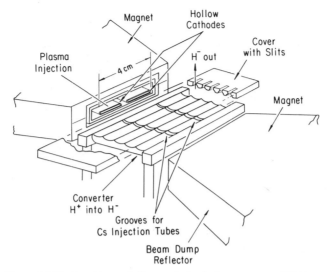

Figure 10.17. Brookhaven hollow cathode discharge source of H⁻ ions.

HCD source is shown in Fig. 10.17. A magnetic field of 0.01–0.02 tesla is maintained parallel to the scalloped surface of the molybdenum converter and perpendicular to the direction of the cylindrical grooves. The discharge is started as an oscillating electron discharge with a tantalum filament opposite the hollow cathodes, not shown to Fig. 10.17. The flattened hollow cathodes through which the H_2 is introduced reach emitting temperature and take over the function of the discharge cathode. The resultant plasma is in the form of a thin sheet close to the converter surface. The slotted cover, shown broken away in Fig. 10.17, serves as the anode and the converter is run at −100 to −150 volts relative to the anode. The discharge was run steady state with the extractor, made of a parallel array of tungsten wires, not shown in Fig. 10.17, pulsed to 7.5 kV with a duty cycle of ≤ 0.2 and pulse lengths up to 1 second. The extracted H⁻ current was 0.2–0.3 A, with approximately equal electron currents, from five slits each 0.2 cm × 5 cm. The neutral gas flow of 0.3 torr-liters/second which was required translates to a gas utilization (see Problem 8.3) of 6% to 9%.

Although this source and the SITEX source are given separate sections because they were approached with different points of view, it is easy to recognize that they are, in fact, highly similar. The ohmically heated SITEX cathode is replaced in the HCD source by the tantalum tubes which are probably capable of yielding greater current density. In addition, the slits in the HCD source are perpendicular to the field, rather than parallel to B, as in SITEX. This was shown by Alessi and Sluyters (1980) to improve output of the magnetron source by a factor of 2. The advantages would appear to be entirely with the hollow cathode discharge. In fact, except for a gas efficiency approximately twice as high as for the SITEX, the performance figures, current density, and ratio of electron current to negative ion current have not come up to those given for SITEX.

G. Multipole Containment Source with a Converter

With the configuration shown in Fig. 10.18, Leung and Ehlers (1982) gave the surface plasma approach to H⁻ formation a new and greater flexibility. The plasma is generated in a chamber bounded by linear magnetic cusps. Of the 14 rows of magnets shown all but the two that bound the extraction aperture are $SmCo_5$. Those two are ceramic magnets so that the field through which the H⁻ ions must pass has a maximum value of only 80 gauss, large enough to constrain electron flow but not large enough to produce an excessive deflection of the ions as they exit the source.

The cathode consisted of eight tungsten filaments, four in each of the two rows indicated in Fig. 10.18. These ran at about 70 volts negative relative to the boundary which serves as the anode and delivered currents as high as 200 A.

The cylindrically contoured converter is made of molybdenum and is 8 cm wide (in the illustrated direction) by 25 cm long (in the direction perpendicular to the figure). The rear and side surfaces are protected from the plasma by a boron nitride housing.

With a 100 A discharge and a converter voltage of −160 A the converter current was about 20 A of positive ions accelerated across the sheath that forms at its surface. With the cesium jet spraying cesium vapor at the surface the resultant

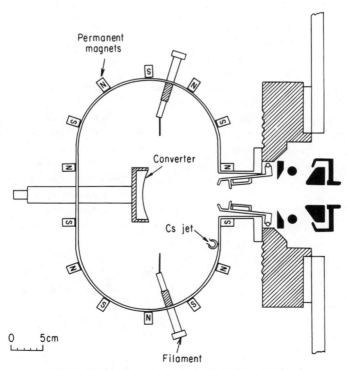

Figure 10.18. H⁻ source of Leung and Ehlers (1982).

steady current of H^- ions that leaves the surface and is accelerated across the sheath and through the 3 cm $\times$ 25 cm exit aperture is 1.1 A.

The converter is located in a region of zero field but the ions move through a crossed magnetic field to reach the aperture. In this region the electrons are much cooler than in the zero field region and the plasma density could be controlled by the potential on the electrically isolated electrode which defines the exit aperture, as discussed in Sec. 8.14. In fact, the plasma density at the exit aperture could be so effectively reduced by a small positive voltage that the positive ion density could be made approximately equal to the density of H^- ions flowing from the converter. The ratio of the electron current to H^- ion current was only 0.038, under conditions of high pumping speed which reduced the gas density in the accelerator gap. Under these conditions the gas utilization was approximately 13%.

The beam was accelerated to 34 keV per ion for a pulse length of 7 seconds but this time was limited not by any source consideration but by the thermal capacity of the beam stop.

10.4 Volume Production of H^-

A. Historical Development

It is clear that negative ions must exist in a plasma and small negative ion currents have been extracted from plasmas for many years. Ehlers (1965) raised the level of achievable currents to 5 mA in a hot filament, oscillating electron source with transverse extraction through a small aperture for cyclotron application. His H^- current density was greater than 40 mA/cm^2.

Another early direction which was followed was the use of the duoplasmatron as a source of negative ions. Lawrence et al. (1965) discovered that the displacement of the extraction aperture in the anode relative to the intermediate electrode increased the H^- yield and dramatically decreased the electron current. With a 0.9 mm diameter aperture displaced about 0.9 mm they obtained 100 μA of H^- ions. Abroyan et al. (1972) with a similar geometry obtained pulsed H^- beams of 2 mA. Sluyters and Prelec (1973) and Prelac and Sluyters (1973) modified the duoplasmatron by introducing a rod on the axis from the cathode region through the intermediate electrode region to within a few millimeters of the anode and restored the extraction aperture to its central position. Using this hollow discharge duoplasmatron Prelec and Sluyters (1973) obtained 8 mA of H^- when the gas was H_2. When cesium was introduced with the H_2, a maximum H^- current of 18 mA was obtained. Similar results were reported by Kobayashi et al. (1976).

When it became clear that H^- beams of the order of amperes would be required for magnetic fusion the production through double charge transfer and then through surface interaction became, in that order, the favored directions. In the 1970s, however, a group at the Ecole Polytechnique at Palaiseau in France began making measurements on the H^- density in hydrogen plasmas with unexpected results. In a low density (10^{10} cm^{-3}) hydrogen plasma Bacal et al. (1977) found H^- densities of the order of 100 times greater than they were able to explain in terms of H^-

creation and destruction processes. The assumptions made in their analysis included the existence in their plasma of a single neutral species, H_2, and a single positive ion species, H_2^+, which are reasonable approximations for the low density plasma with which they worked. At the same conference, Crandall and Barnett (1977) presented an extremely thorough review of the known processes for production and destruction of H^- and concluded that the cross sections for destructive processes were so much larger than those for production of H^- that the production of high densities of H^- in laboratory plasmas was not likely.

Three years later, at the next major negative ion beam conference, Bacal et al. (1980) had extended their plasma studies to a density of 10^{11} cm^{-3} and found negative ion fractions n_{H^-}/n_e as high as 0.35, but the program of this conference did not contain a single report of the production of a beam of H^- ions in which the ions were produced within a discharge plasma. It is clear that substantial advantage could accrue from the extraction of H^- ions directly from the volume of a hydrogen plasma, in that no cesium or other contaminant vapor was involved, as in double charge transfer or surface plasma sources. However, there is a fundamental difficulty in the extraction of negative ions from a plasma. Because a plasma is normally positive relative to its boundaries the plasma is a trap for negative ions. This difficulty cannot be overcome simply be making the extraction aperture more positive than the anode. Without a magnetic field to prevent electrons from getting to the positive electrode the plasma will go positive relative to the most positive electrode on its boundary.

The extraction problem was attacked by Leung et al. (1983) through an application of the tandem cusp configuration described in Sec. 8.14 and shown in this application in Fig. 10.19. This source also provided evidence for a mechanism of H^- formation that could explain the large densities found by the Palaiseau group.

The negative ion extraction studies with the tandem plasma configuration were carried out in three stages. In the first step there was no magnetic filter, that is,

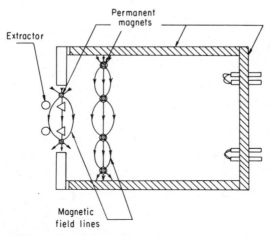

Figure 10.19. Schematic of H^- ion source of Leung, Ehlers, and Bacal (1983).

the permanent magnets creating the transverse field were absent. The extracted H^- currents were minute (2 μA) and the observed electron currents approximately 9000 times as large. When the magnetic filter was introduced, separating the plasma into two portions, with the extractor side plasma having a much lower electron temperature, the H^- current rose to about 10 μA and the electron current dropped substantially. A bias of +2.5 volts on the plasma electrode (the electrode facing the plasma and containing the extraction aperture) increased the H^- current to 23 μA but the electron current was still 100 times greater. When the transverse field was placed across the extractor, as shown in Fig. 10.19, the electron current was reduced to the level where electron and H^- currents were about equal. The H^- yield increased approximately linearly with discharge current and at 350 A Leung and Ehlers (1983) were able to obtain a current density of 38 mA/cm^2 from a 0.15 cm $\times$ 1.3 cm slot.

The work with tandem multicusp plasmas provided the verification of a growing awareness of a mechanism, described in the next section, which could account for the large H^- densities that had been observed, and set the stage for the development of H^- volume sources yielding beams of interest to the fusion effort. By the time of the two major negative ion conferences in 1986 (Palaiseau, March 5–7 and Brookhaven, October 27–31) a majority of the papers were related to volume production. The nature of some of the sources presented will be described after a discussion of the fundamental processes.

B. Fundamental Processes

When Bacal et al. (1977) found that their H^- densities were 100 times greater than anticipated, the expected levels were based on the following processes of H^- formation and destruction:

dissociative attachment:

$$e + H_2 \rightarrow H^- + H,$$

collisional dissociation:

$$e + H_2 \rightarrow H^- + H^+ + e,$$

dissociative recombination:

$$e + H_2^+ \rightarrow H^- + H^+,$$

collisional detachment:

$$e + H^- \rightarrow H + 2e,$$

and recombination:

$$H^- + H_2^+ \rightarrow neutrals.$$

The first three of these reactions, those leading to H^- formation, all had very small cross sections, whereas the collisional and recombination reactions by which H^- ions were destroyed had large cross sections. It is clear that these reactions alone cannot lead to the large H^- fractions which have been observed in hydrogen plasmas. Crandall and Barnett (1977) did a more extensive examination of reactions leading to the formation and destruction of H^-. In the former category they examined the following reactions which were not considered by Bacal et al. because they had assumed only a single neutral species, H_2, and a single positive ion species H_2^+:

radiative capture:

$$e + H \rightarrow H^- + h\nu$$

and three body capture:

$$e + H + t \rightarrow H^- + t,$$

where t stands for the third object, plus the reactions

$$H + H \rightarrow H^- + \cdots,$$
$$H + H_2 \rightarrow H^- + \cdots,$$
$$H^+ + H_2 \rightarrow H^- + \cdots,$$

and

$$H_2^+ + H_2 \rightarrow H^- + \cdots.$$

The cross section for the radiative capture is minuscule ($\sim 5 \times 10^{-22}$ cm^2). The three body capture would require densities much too high ($\sim 10^{18}$ cm^{-3}) to equal radiative attachment rates. The last four reactions listed have reasonable cross sections for an incident particle of energy of the order of 10 keV and are much too small to contribute at the energies which exist in a laboratory plasma. On the other hand, Crandall and Barnett considered, in addition to the destructive processes considered by Bacal et al. (1977), photodetachment,

$$H^- + h\nu \rightarrow H + e,$$

and the electron transfer reaction,

$$H^- + H^+ \rightarrow H + H.$$

Both of these reactions would add significantly to the destruction of H^-. They conclude the production of high densities of H^- in hydrogen plasmas is not likely.

However, the results of Bacal and co-workers were not to be denied, especially when their measurements were augmented by the photodetachment diagnostic experiments of Hamilton et al. (1980). The reaction which was finally discovered to have a cross section large enough to explain the observed H^- densities was the dissociative attachment reaction,

$$e + H_2(v^*) \rightarrow H_2^- \rightarrow H^- + H,$$

where $H_2(v^*)$ stands for a vibrationally excited H_2 molecule. (These are often referred to as ro-vibrational molecules since rotational states may be excited as well.) For the lowest electronic state, with the vibrational quantum number 6 or higher, the cross sections were calculated by Crandall and Meyer (1980) to be larger than the dissociative attachment cross section for H_2 in the ground state by a factor of 10^4 to 10^5. In addition, it was clear that large amounts of vibrationally excited H_2 could be expected through two different paths. In the first of these, electrons of energy >20 eV can produce the reaction

$$e + H_2 \rightarrow H_2^*,$$

where H_2^* stands for electronically excited H_2. The electronically excited hydrogen then decays radiatively to the vibrationally excited ground electronic state. Another path which leads to $H_2 (v^*)$ is the reaction

$$e + H_3^+ \rightarrow H_2 (v^*) + H.$$

H_3^+ normally exists in a laboratory plasma in high density due to the reaction

$$H_2^+ + H_2 \rightarrow H_3^+ + H.$$

I have omitted reactions leading to $H_2(v^*)$ which involve the destruction of H^- ions.

The reason for the effectiveness of the source of Leung et al. (1983) in which the main plasma, containing a high density of energetic primaries, is separated from the extraction plasma containing only very low energy electrons ($T_e \approx 1$ eV) becomes clear in terms of H^- production through dissociative attachment to vibrationally excited molecules. In the main plasma copious amounts of high energy (> 20 eV) electrons lead to a large rate of production of vibrationally excited hydrogen molecules which pass freely through the filter. On the extraction side of the filter the dissociative attachment reaction which produces the desired H^- ions occurs rapidly in the flux of very low energy electrons in this region. At the same time the very low electron temperature will virtually eliminate the collisional detachment of electrons from the H^- ions, although electron transfer and recombination reactions will still occur. It is clear there must be an optimum thickness to the extraction side plasma.

C.   Tandem Multicusp Plasma Sources

Leung, Ehlers and the group at Lawrence Berkeley Laboratory continued to optimize the tandem plasma source described in the previous section. Leung et al. (1985)

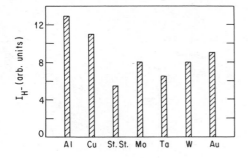

Figure 10.20. Effect of various wall materials on H⁻ currents.

found a sizable increase in extractable H⁻ current if the extractor was placed very near the filter and studied the effect of wall material. Their results are displayed in Fig. 10.20. Clearly Al and Cu are the preferred materials with Cu being twice as good as the frequently used stainless steel. They attribute their result to the effect of secondary emission under 90 eV electron bombardment. Considering the emission energy and the plasma-to-wall potential secondary emission will introduce electrons with approximately 8 eV energy into the plasma. In a separate experiment it was found that 8 eV electrons reduce the H⁻ yield substantially. This plus the inverse correlation with secondary emission yields led to their conclusion.

A wall effect can be expected for another reason, apparently not considered. The number of wall collisions that H_2 $(v*)$ is able to survive will surely depend on the wall material. It is possible that an optimized source will have one material at the cusps, where electron bombardment will be important, and another between cusps, where neutral particles strike.

In a small (7.5 cm diameter by 8 cm long) optimized tandem source Ehlers et al. (1986), operating with a 110 A, 150 volt discharge, achieved a current density of 240 mA/cm².

The record for total current with a tandem multicusp source is that achieved by Okumura et al. (1986), whose source is shown in Fig. 10.21. Their plasma generator was 21 cm × 36 cm by 15 cm deep and Fig. 10.21 shows the 21 cm × 15 cm cross section. The plasma is surrounded by three transverse rows of $SmCo_5$ magnets on each side wall and five rows on the end plate of the chamber. This is certainly not a distribution of magnets that will yield a zero longitudinal field along the center line, but the authors make no comment about the effects of the residual field in the plasma. The magnetic filter consisted of five water-cooled tubes in which small $SmCo_5$ magnets are inserted and the filter strength ($\int B \, dl$) was about 0.2 tesla-meter. Actually their best performance was obtained when the filter was weakened by a removal of the two rows of magnets on either side of the central row.

The extraction and acceleration of the H⁻ ions were accomplished by four grids, each with 209 apertures of 9 mm diameter within a rectangular area 12 cm × 26 cm. A cross section through a single aperture is shown in Fig. 10.22 together with the applied voltages. The extraction grid had rows of tiny magnets, magnetized as shown, producing a zero intergrated transverse magnetic field. This field deflected

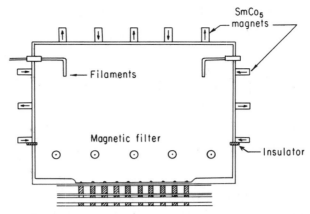

Figure 10.21. Schematic cross section through the smaller dimension of the multiaperture extractor tandem plasma source of Okumura et al. (1986).

all the extracted electrons into the extraction grid, and the suppressor grid served to keep the secondary electrons from being accelerated through the full voltage. With the weakened magnetic filter and a discharge current of 700 A at 70 volts, a peak H⁻ beam of 1.26 A was extracted for a pulse length of 0.2 second. The optimum pressure in the discharge was about 1 Pa (7.5×10^{-3} torr) for optimum performance at the high current levels.

D.   Variations on the Tandem Plasma Source

A brief restatement of the mechanisms by which H⁻ ions are made within a plasma volume may be helpful. The first step is the generation of H_2 molecules in high vibrational states. The most effective way of doing this is by bombardment of ground state H_2 molecules by energetic ($kT > 20$ eV) electrons which excite electronic states of the molecule. These quickly decay to relatively long lived high vibrational levels of the ground electronic state.

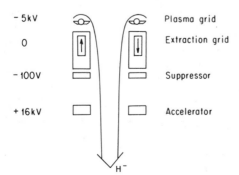

Figure 10.22. Extraction and acceleration electrodes of the source of Okumura et al. (1986).

These molecules move readily through the filter magnetic field, which effectively blocks out the fast electrons. The low energy ($kT < 1$ eV) electrons readily produce a dissociative attachment reaction with these vibrationally excited molecules leading to H⁻ ions and H atoms. Without the energetic electrons, collisional detachment of the bound electron from H⁻ cannot occur, although H⁻ ions can still be lost by recombination or charge transfer with positive ions.

A Radial Tandem Plasma Source. Various configurations, other than the tandem multicusp plasma device, as described in the previous section, have been tried. Bacal et al. (1986) built the source shown in Fig. 10.23. In this source the filaments, which are the cathodes of the discharge, are placed around the periphery of the source, between the rows of magnets which form linear cusps along the cylindrical boundary of the source. The electrons in the plasma in the center of the chamber are those that have migrated across the magnetic fields which are peripheral to this region and are therefore cold electrons. Bacal et al. essentially created a cylindrical version of the planar tandem plasma device.

However, the planar configuration has the advantage that the magnetic filter could be positioned arbitrarily close to the extraction plane. This means that, in the competition between dissociative attachment, which creates H⁻, and the processes which consume H⁻, the extraction plane is at the position where the H⁻ current density is a maximum. In the radial arrangement of Bacal et al., H⁻ ions formed near the rear of the source have to run a substantial gauntlet before reaching the extraction aperture. Bacal et al. obtained an H⁻ current of 0.5 mA from a 0.8 cm diameter aperture, accompanied by an electron current of 2 mA which was swept to the extraction electrode by the pair of magnets seen in Fig. 10.23. As noted in other sources a potential of +2 volts on the plasma electrode increased

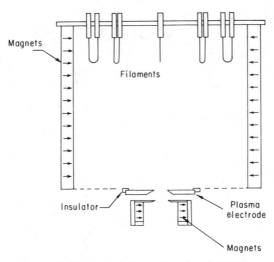

Figure 10.23. H⁻ ion source of Bacal et al. (1986).

the H⁻ current and greatly decreased the electron current. The values given above are those obtained with this bias.

Cusped Plasma Containment with a Transverse Field. The tandem plasma device is so named because in its original concept there were two plasma regions with very different electron temperatures separated by a magnetic filter. What has been found, however, is that the optimum position of the filter is so close to the plasma electrode of the extraction system that there is no uniform plasma region on the plasma electrode side of the filter. Rather, there is an electron temperature which gradually decreases with distance toward the plasma electrode. At the same time the flow of H⁻ ions should gradually increase with distance and, for the correct $\int Bdl$, I_H^- should be a maximum at the plasma electrode. This suggests that it may be unnecessary to have the traverse magnetic filter field confined to a narrow sheet, an arrangement which required water-cooled columns of magnets immersed in the plasma region.

The Culham Laboratory approach, described, for example, the McAdams et al. (1986), achieved a broad transverse field by having two rows of aligned linear magnets on each of two sides of a rectangular plasma containment device (Fig. 10.24), with the remaining portions of the surface, other than the unmagnetized

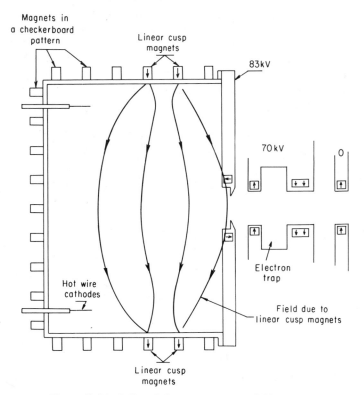

Figure 10.24. Culham Laboratory transverse field source.

extraction plane, covered with a checkerboard array of magnets, roughly as shown in Fig. 8.26. With the arrangement shown there a current of 145 mA at 83 keV was extracted from a 2.4 cm diameter aperture. Approximately five times as many electrons are extracted but these are filtered out by the magnetic fields at the second electrode. Like other experimenters, the Culham group was able to greatly reduce the ratio of extracted electron to H⁻ current with a very small potential (2.5 volts) on the plasma electrode. However, unlike others, they found that this also produced a significant reduction in the H⁻ current.

Penning-Type Source without Cesium.   Although the early work of Ehlers (1965) yielded an H⁻ current density of 40 mA/cm² from a small hole in the anode of an oscillating electron source, further pursuit of this configuration was neglected until the work of Jimbo et al. (1986), who, in a similar geometry, achieved a current density of 100 mA/cm² with a total H⁻ current of 9.7 mA. The source used by them is shown in Fig. 10.25. In this source the plasma column which faces only cathode surfaces along magnetic field lines contains primary as well as plasma electrons. However, fast electrons cannot readily penetrate the region between this column and the extraction aperture, and those that do are readily swept out to the anode surface. The field in this latter region then plays the role of a magnetic filter and the plasma electrons will become cooler as the aperture is approached. As in other volume sources H_2 molecules get excited to high vibrational states in the region containing energetic electrons and these vibrationally excited molecules undergo dissociative attachment in the region containing only low energy electrons.

Generically, the source described by Leung et al. (1986), shown in Fig. 10.26, is the same. The LaB_6 cathodes are not ohmically heated but, once struck, the discharge can heat the cathodes to thermionic emission temperatures. The anode ribs shown in the figure provide the magnetic filter region with the low electron temperature plasma which is desired to optimize the dissociative attachment and minimize collisional detachment. H⁻ current densities of 350 mA/cm² have been obtained for arc currents of 55 A. Removal of the anode ribs reduces the extracted H⁻ current by a factor of 3.

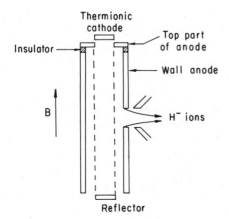

Figure 10.25. Oscillating electron source of Jimbo et al. (1986).

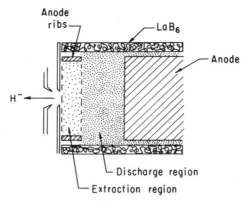

Figure 10.26. Penning-type of source due to Leung et al. (1986).

The pulse length of this source was limited to several hundred microseconds by heating considerations.

The performance of this source without cesium, with H⁻ ions almost certainly orginating in the volume of the plasma, must cast a little doubt on the interpretation of the Dudnikov–Penning source of H⁻ ions. There is little doubt that the addition of cesium to the Dudnikov source greatly increased the extracted H⁻ current, but it becomes necessary to examine the possibility that this is also a volume effect, related to changes in electron energy distribution, rather than a surface effect.

E. Some Speculative Sources of H⁻

In the contex of the currently envisioned mechanism of volume production of negative ions, that is, vibrational excitation of H_2 in a plasma containing energetic electrons followed by dissociative attachment in a very low electron temperature plasma, the single-ring magnetic cusp ion source of Brainard and O'Hagan (1983), described in Sec. 8.3E, should be recalled. Although the configuration is very different from those which have been successful as plasma sources of H⁻, the Brainard–O'Hagan source is characterized by a plasma containing energetic electrons at the rear and a low electron temperature plasma at the ion extraction aperture. It ought to be examined for its negative ion generating capability.

Since a magnetic mirror leads to a selective rejection of the hot electrons as the mirror is approached, another type of magnetic filter may be worth trying: the picket fence boundary of Fig. 8.24 shown embodied in the proposed source of Fig. 10.27. Permanent magnets in the arrangement shown in Fig. 8.25c produce an array of similar cusps, but the slow electrons which get through the cusps will in large measure strike the magnet faces instead of penetrating to the opposite side.

A variant of this scheme might have the picket fence outside the plasma electrode, constituting the first electrode of the accelerating system. In this case the arrangement of permanent magnets as in Fig. 8.25c might actually be advantageous over the picket fence.

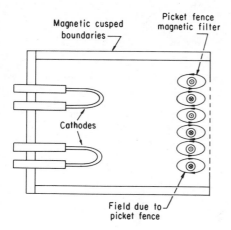

Magnetic cusped boundaries

Picket fence magnetic filter

Cathodes

Field due to picket fence

Figure 10.27. A proposed H⁻ source geometry.

10.5 Negative Ion Extraction and Acceleration

As mentioned briefly in Sec. 10.4A the extraction of negative ions from the volume of a plasma is a much more complicated problem than the extraction of positive ions, which is discussed in Chapter 5. In the positive ion case one is normally unconcerned with the fact that the plasma may contain negative ions or even energetic electrons. It is a good approximation to consider the plasma composed of maxwellian electrons of temperature T and the positive ions. The latter drift to the boundaries of the plasma with energies of the order of kT. Where the ions drift to an extraction aperture the shape of the surface defining the boundary between the near neutral plasma and the positive ion beam is determined by well defined criteria, even though the computation of the shape may pose some difficulties.

The negative ion extraction situation is much more complicated. In the volume production sources favorable results are associated with a very low electron temperature near the extraction aperture. In this case, even if the plasma is positive relative to its boundaries, it is only slightly so and the negative ions, which are formed by a dissociative attachment reaction with vibrationally excited molecules, can have enough energy to surmount the barrier. In some cases the effective trapping of electrons by magnetic fields which bound the plasma, including a magnetic field across the extraction aperture, may cause the plasma to go negative relative to its boundary and actually accelerate negative ions out of the plasma. In any case the extraction of negative ions is always accompanied by a flow of electrons which must be taken into account in any space charge analysis. A thorough review of computer modeling of negative ion beam formation is presented by J. H. Whealton (1986). Further details of the extraction and acceleration of negative ions are outside the scope of this book.

References

M. Abele and W. Meckbach, Design and Performance of a Hot Cathode Magnetically Collimated Arc Discharge Ion Source, *Rev. Sci. Instrum.* **30,** 335 (1959).

M. E. Abdalaziz and A. M. Ghander, High Current P.I.G. Ion Source at Low Gas Pressure, *IEEE Trans. Nucl. Sci.* **NS-14,** No. 3, p. 53, June 1967.

M. Abramowitz and I. A. Stegun, *Handbook of Mathematical Functions,* National Bureau of Standards, Applied Mathematics Series, Vol. 55, 319 (1964).

M. A. Abroyan, G. A. Nalivaiko, and S. G. Tsepakin, Pulsed Source of Negative Hydrogen Ions, *Sov. Phys. Tech. Phys.* **17,** 690 (1972).

J. Alessi and Th. Sluyters, Regular and Asymmetric Negative Ion Magnetron Sources with Grooved Cathodes, Brookhaven 1980, p. 153.*

J. G. Alessi, A. Hershcovitch, and Th. Sluyters, Cesiated Porous Mo Converter for Intense Negative Ion Sources, *Rev. Sci. Instrum.* **55,** 8 (1984).

P. W. Allison, A Direct Extraction H^- Ion Source, *IEEE Trans. Nucl. Sci.* **NS-24,** 1594 (1977).

P. Allison, H. V. Smith, Jr., and J. D. Sherman, H^- Ion Source Research at Los Alamos, Brookhaven 1980, p. 171.

O. A. Anderson, *A Compact High-Energy Neutral Beam System,* Lawrence Livermore Laboratory Report UCID-16914, Aug. 15, 1975.

O. A. Anderson, *Parasitic Components from Charge Transfer in Neutral Beams for Fusion,* Lawrence Livermore Report UC1D-17581, Feb. 1, 1978.

O. A. Anderson and E. B. Hooper, Jr., Plasma Production and Flow in Negative Ion Beams, Brookhaven 1977, p. 205.

*Brookhaven 1977, 1980, 1983, and 1986 refer to Proceedings of the International Symposia on the Production and Neutralization of Hydrogen Ions and Beams, Brookhaven National Laboratory, Upton, New York, on Sept. 26–30, 1977, Oct. 6–10, 1980, Nov. 14–18, 1983, and Oct. 27–31, 1986, respectively. The 1983 proceedings have been published in hardcover as AIP Conference Proceedings, Number 111, ed. by K. Prelec.

M. von Ardenne, New Developments in Applied Ion and Nuclear Physics, *Atomkernenergie* **1**, 121 (1956).

M. Bacal, E. Nicolopoulou, and H. J. Doucet, Production of Negative Hydrogen Ions in Low Pressure Hydrogen Plasmas, Brookhaven 1977, p. 26.

M. Bacal, A. M. Bruneteau, H. J. Doucet, W. G. Graham, and G. W. Hamilton, H^- and D^- Production in Plasmas, Brookhaven 1980, p. 95.

M. Bacal, H. J. Doucet, G. Labaune, H. Lamain, C. Jacquot, and S. Verney, Cesium Supersonic Jet for D^- Production by Double Electron Capture, *Rev. Sci. Instrum.* **53**, 159 (1982).

M. Bacal, F. Hillion, M. Nachman, and W. Steckelmacher, Progress in Developing a "Volume" Hydrogen Negative Ion Source, Brookhaven 1983, p. 418.

M. Bacal, J. Bruneteau, P. Devynck, and F. Hillion, Volume Production of H^- Ions at Ecole Polytechnique. A Method for Extracting Volume Produced Negative Ions, Brookhaven 1986.

J. Backus, Theory and Operation of a Phillips Ionization Gauge Type Discharge, in Guthrie and Wakerling (1949), Chap. 11.

C. Bailey, D. L. Drukey, and F. Oppenheimer, A Magnetic Ion Source, *Rev. Sci. Instrum.* **20**, 189 (1949).

B. Banks, Composite Ion Accelerator Grids, Proc. 3rd Int. Conf. on Electron and Ion Beam Science and Tech., Boston, Mass., (1968).

C. F. Barnett, J. A. Ray, E. Ricci, M. I. Wilker, E. W. McDaniel, E. W. Thomas, and H. B. Gilbody, *Atomic Data for Controlled Fusion Research,* ORNL-5206 (vol. I) and ORNL-5207 (vol. II), Oak Ridge National Lab., Oak Ridge, TN (1977).

D. G. Bate, *A Computer Program for the Design of High Perveance Ion Sources,* Culham Laboratory Report CLM-R53 (1966), Abingdon, Oxon, England.

T. R. Bates and A. T. Forrester, Coupled Molecular Flow and Surface Diffusion: Application to Cesium Transport, *J. Appl. Phys.* **38**, 1956 (1967).

R. J. Bearman and F. H. Horne, Comparison of Theories of Heat of Transport and Thermal Diffusion with Experiments on the Cyclohexane–CCl_4 System, *J. Chem. Phys.* **42**, 2015 (1965).

R. T. Bechtel, Discharge Chamber Optimization of the SERT II Thruster, *J. Spacecraft Rockets* **5**, 795 (1968).

R. T. Bechtel, B. A. Banks, and T. W. Reynolds, *Effect of Facility Backscattered Material on Performance of Glass Coated Accelerator Grids for Kaufman Thrusters,* AIAA 9th Aerospace Science Meeting, New York, Jan. 25–27, 1971, paper no. 71-156.

Yu. I. Bel'chenko and G. I. Dimov, Pulsed Multiampere Source of Negative Hydrogen Ions, Brookhaven 1983, p. 363.

Yu. I. Bel'chenko and V. G. Dudnikov, Negative Ion Production in Surface-Plasma Sources with Unclosed Electrode Drift Discharges, *J. Phys. (Paris)* **40**, C7-501 (1979).

Yu. I. Bel'chenko, G. I. Dimov, and V. G. Dudnikov, Emission of Intense Negative-Ion Fluxes from a Surface Bombarded by Fast Particles from a Discharge, *Izv. Akad. Nauk. SSSR, Ser. Fiz.* **37**, 2573 (1973). In English: *Bull. Acad. Sci. USSR, Phys. Ser.* **37**, (12), 91 (1973).

Yu. I. Bel'chenko, G. I. Dimov, V. G. Dudnikov, and A. A. Ivanov, Formation of Negative Ions in a Gas Discharge, *Dokl. Akad. Nauk. SSSR* **213**, 1283 (1973). In English: *Sov. Phys. Dokl.* **18** (12), 814 (1974).

Yu. I. Bel'chenko, G. I. Dimov, and V. G. Dudnikov, *Surface-Plasma Source of Negative Ions,* Proc. 2nd Symp. Ion Sources and Formation of Ion Beams, Berkeley, Oct. 22–25, 1974, paper VIII-1.

Yu. I. Bel'chenko, G. I. Dimov, and V. G. Dudnikov, Physical Principles of the Surface Plasma Method for Producing Beams of Negative Ions, Brookhaven 1977, p. 79.

D. Bohm, Qualitative Description of the Arc Plasma in a Magnetic Field, in Guthrie and Wakerling (1949), Chap. 1.

D. Bohm, Minimum Ionic Kinetic Energy for a Stable Sheath, in Guthrie and Wakerling (1949), Chap. 3.

D. Bohm, E. H. S. Burhop, H. S. W. Massey, and R. M. Williams, A Study of the Arc Plasma, in Guthrie and Wakerling (1949), Chap. 9.

J. P. Brainard and J. B. O'Hagan, Single Ring Magnetic Cusp Ion Source, *Rev. Sci. Instrum.* **54,** 1497 (1983).

G. R. Brewer, *Ion Propulsion Technology and Applications,* Gordon and Breach, New York (1970).

S. C. Brown, *Introduction to Electrical Discharges in Gases,* Wiley, New York (1966).

S. C. Brown, *Basic Data of Plasma Physics, 1966,* 2nd ed., rev., M.I.T. Press, Cambridge (1967).

C. Buxbaum, The LM Cathode, A New Thermionic High-Performance Metal Cathode, *Brown-Boveri Rev,* p. 43 (1979).

C. Buxbaum and G. Gessinger, Reaction Cathode, United States Patent No. 4,019,018, April 19, 1977.

C. E. Carlston, G. D. Magnuson, A. Comeaux, and P. Mahadevan, Effect of Elevated Temperature on Sputtering Yields, *Phys. Rev. A* **138,** 759 (1965).

A. Caruso and A. Cavaliere, The Structure of the Collisionless Plasma-Sheath Transition, *Nuovo Cimento* **26,** 1389 (1962).

S. Chandrasekhar, Time of Relaxation of Stellar Systems, *Astrophys. J.* **93,** 285 (1941).

S. Chandrasekhar, *Principles of Stellar Dynamics,* University of Chicago Press, Chicago (1942), Chap. 2 and Sec. 5.6.

S. Chandrasekhar, Dynamical Friction. I. General Considerations: The Coefficient of Dynamical Friction, *Astrophys. J.* **97,** 255 (1943).

F. F. Chen, Electrostatic Stability of a Collisionless Plane Discharge, *Nuovo Cimento* **26,** 698 (1962).

F. F. Chen, *Introduction to Plasma Physics,* Plenum, New York (1974).

C. D. Child, Discharge from Hot CaO, *Phys. Rev.* **32,** 492 (1911).

J. D. Cobine, *Gaseous Conductors,* Dover, New York (1958). (Originally published in 1941.)

W. S. Cooper, K. H. Berkner, and R. V. Pyle, Multiple-Aperture Extractor Design for Producing Intense Ion and Neutral Beams, *Nucl. Fusion* **12,** 263 (1972a).

W. S. Cooper, K. H. Berkner, and R. V. Pyle, Production of Intense Ion and Neutral Beams with a Multiple-Slot, Large-Area Extractor, 2nd Int. Conf. Ion Sources, Vienna, Austria, Sept. 11–16, 1972, paper LBL-916.

M. J. Copley and T. E. Phipps, The Surface Ionization of Potassium on Tungsten, *Phys. Rev.* **48,** 960 (1935).

J. R. Coupland, T. S. Green, D. P. Hammond, and A. C. Riviere, A Study of the Ion Beam Intensity and Divergence Obtained from a Single Aperture Three Electrode Extraction System, *Rev. Sci. Instrum.* **44,** 1258 (1973).

D. H. Crandall and C. F. Barnett, Fundamental Atomic Collisional Processes in Negative Ion Sources for H⁻, Brookhaven 1977, p. 3.

D. H. Crandall and F. W. Meyer, Hydrogen Negative Ions and Collisions of Atomic Particles, Brookhaven 1980, p. 1.

F. W. Crawford and A. B. Cannara, *Structure of the Double-Sheath in a Hot-Cathode Plasma,* Stanford University Microwave Laboratory Report No. 1261, Nov. 1964.

J. T. Crow, *Space Charge Effects in Ion Beams,* Doctoral dissertation, University of California Los Angeles (1977).

J. T. Crow, A. T. Forrester, and D. M. Goebel, High Performance Low Energy Ion Source, *IEEE Trans. Plasma Sci.* **PS-6,** 535 (1978).

W. K. Dagenhart, W. L. Stirling, H. H. Haselton, G. G. Kelley, J. Kim, C. C. Tsui, and J. H. Whealton, Modified Calutron Negative Ion Source Operation and Future Plans, Brookhaven 1980, p. 217.

W. K. Dagenhart, W. L. Gardner, G. G. Kelley, W. L. Stirling, and J. H. Whealton, SITEX Negative Ion Source Sealing Studies to Produce 200-keV, 10A, Long Pulse D⁻ Beams, Gatlinburg 1981, p. 317.*

W. K. Dagenhart, W. L. Stirling, G. M. Banec, G. C. Barber, N. S. Ponte, and J. H. Whealton, Short-Pulse Operation with the SITEX Negative Ion Source, Brookhaven 1983, p. 353.

H. L. Daley, J. Perel, and R. H. Vernon, Indium Ion Source, *Rev. Sci. Instrum.* **37,** 473 (1966).

S. Datz and E. H. Taylor, Ionization on Platinum and Tungsten Surfaces: I. The Alkali Metals, *J. Chem. Phys.* **25,** 389 (1956).

R. C. Davis, O. B. Morgan, L. D. Stewart, and W. L. Stirling, A Multiampere Duo-PIGatron Ion Source, *Rev. Sci. Instrum.* **43,** 278 (1972).

R. C. Davis, T. C. Jernigan, O. B. Morgan, L. D. Stewart, and W. L. Stirling, Duo-PIGatron II Ion Source, *Rev. Sci. Instrum.* **46,** 576 (1975).

D. P. DeBruijn, J. Neuteboom, V. Sidis, and J. Los, A Detailed Experimental Study of the Dissociative Charge Exchange of H_2^+ with Ar, Mg, Na and Cs Targets at keV Energies, *Chem. Phys.* **85,** 215 (1984).

M. Delaunay, R. Geller, C. Jacquot, P. Ludwig, P. Sermet, J. C. Rocco, F. Zadworny, J. B. Bergstrom, G. Hellblom, R. Pauli, and H. Wilhelmsson, Large Negative Ion Source for Energetic Neutral Beams, Brookhaven 1983, p. 438.

J.-L. Delcroix and A. R. Trindade, Hollow Cathode Arcs, *Adv. Electron. Electron Phys.* **35,** 87 (1974).

R. A. Demirkhanov, Yu. V. Kursanov, and V. M. Blagoveshchenskii, High-Intensity Proton Source, *Prib. Tekh. Eksp.,* No. 1, pp. 30–33, Jan.–Feb. 1964. In English: *Instrum. Exp. Tech.* Pt. 1, 25 (1964).

G. I. Dimov and G. V. Roslyakov, Conversion of a Beam of Negative Hydrogen Ions to Atomic Hydrogen in a Plasma Target at Energies between 0.5 and 1 MeV, *Nucl. Fusion* **15,** 551 (1975).

V. G. Dudnikov, Surface Plasma Source of Negative Ions with Penning Geometry, 4th USSR National Conference of Particle Accelerators, 1974, translation LA-TR-75-4, Los Alamos NM (1975).

V. G. Dudnikov, Some Effects of Surface-Plasma Mechanism for Production of Negative Ions, Brookhaven 1980, p. 137.

S. Dushman, Electron Emission from Metals as a Function of Temperature, *Phys. Rev.* **21,** 623 (1923).

S. Dushman, Thermionic Emission, *Rev. Mod. Phys.* **2,** 381 (1930).

---

*Gatlinburg 1981 refers to the Proceedings of the Third Neutral Beam Heating Workshop, Gatlinburg, Tennessee, Oct. 19–23, 1981.

K. W. Ehlers, Design Considerations for High-Intensity Negative Ion Sources, *Nucl. Instrum. Methods* **32,** 309 (1965).

K. W. Ehlers and K. N. Leung, Characteristics of the Berkeley Multicusp Ion Source, *Rev. Sci. Instrum.* **50,** 1353 (1979).

K. W. Ehlers and K. N. Leung, Effect of a Magnetic Filter on Hydrogen Ion Species in a Multicusp Ion Source, *Rev. Sci. Instrum.* **52,** 1452 (1981).

K. W. Ehlers and K. N. Leung, Further Study on a Magnetically Filtered Multicusp Ion Source, *Rev. Sci. Instrum.* **53,** 1423 (1982).

K. W. Ehlers, W. R. Baker, K. H. Berkner, W. S. Cooper, W. B. Kunkel, R. V. Pyle, and J. W. Stearns, Design and Operation of an Intense Neutral Beam Source, *J. Vac. Sci. Technol.* **10,** 922 (1973).

K. W. Ehlers, K. N. Leung, R. V. Pyle, and W. B. Kunkel, Characteristics of a Small Multicusp H⁻ Source, Brookhaven 1986.

M. P. Ernstene, A. T. Forrester, R. C. Speiser, and R. M. Worlock, Multiple Beam Ion Motors, Presented at ARS conference in 1960, reproduced in Langmuir et al. (1961).

M. P. Ernstene, E. L. James, G. W. Purmal, R. M. Worlock, and A. T. Forrester, Surface Ionization Engine Development, *J. Spacecraft Rockets* **3,** 744 (1966).

M. P. Ernstene, A. T. Forrester, R. C. Speiser, G. Sohl, A. H. Firestone, and P. O. Johnson, Gas Discharge Neutralizer Including a Charged Particle Source, U.S. Pat. 3,523,210, Aug. 4, 1970.

C. E. Fay, A. L. Samuel, and W. Shockley, On the Theory of Space Charge between Parallel Plane Electrodes, *Bell System Tech. J.* **17,** 49 (1938).

J. H. Fink and G. W. Hamilton, A Neutral-Beam Injector for the Tandem-Mirror Fusion Reactor Delivering 147 MW of 12-MeV D°, *IEEE Trans. Plasma Sci.* **PS-6,** 417 (1978).

J. H. Fink and B. W. Schumacher, The Anatomy of a Thermionic Electron Beam, *Optik* **39,** 543 (1974).

J. H. Fink and B. W. Schumacher, Characterization of Charged Particle Beam Sources, *Nucl. Instrum. Methods* **130,** 353 (1975).

A. T. Finkelstein, A High Efficiency Ion Source, *Rev. Sci. Instrum.* **11,** 94 (1940).

J. Fletcher and I. R. Cowling, Electron Impact Ionization of Neon and Argon, *J. Phys. B* **6,** L-258 (1973).

A. T. Forrester, Analysis of the Ionization of Cesium in Tungsten Capillaries, *J. Chem. Phys.* **42,** 972 (1965).

A. T. Forrester, Analysis of a Magnetic Double Sheath for Ion Beam Acceleration and Deceleration, *J. Appl. Phys.* **46,** 2051 (1975).

A. T. Forrester, Solution of Space Charge Problems by Conservation of Momentum, *Am. J. Phys.* **50,** 645 (1981).

A. T. Forrester and J. Busnardo-Neto, Magnetic Fields for Surface Containment of Plasmas, *J. Appl. Phys.* **47,** 3935 (1976).

A. T. Forrester and J. M. Dawson, Neutral Beam Line Improvements Resulting from a Reduction of Gas Flow, *IEEE Trans. Plasma Sci.* **PS-6,** 574 (1978).

A. T. Forrester and G. Kuskevics, Editors, *Ion Propulsion,* AIAA Selected Reprint Series, Vol. III American Institute of Aeronautics and Astronautics, New York (1968).

A. T. Forrester and Y. Zlotin, Alternative Configurations of Permanent Magnets, *Bull. Am. Phys. Soc.,* series II, **21,** 1049 (1976).

A. T. Forrester, G. Kuskevics, and B. Marchant, Ionization of Cesium on Porous Tungsten, *Proc. 14th Int. Astronautical Cong.,* Paris, 1963, p. 447.

A. T. Forrester, J. T. Crow, N. A. Massie, and D. M. Goebel, A Multipole Containment-Single Grid Extraction Ion Source, UCLA report PPG-224, June 1975.

A. T. Forrester, D. M. Goebel, and J. T. Crow, IBIS: A Hollow-Cathode Multipole Boundary Ion Source, *Appl. Phys. Lett.* **33,** 11 (1978).

V. V. Fosnight, T. R. Dillon, and G. Sohl, Thrust Vectoring of Multiaperture Cesium Electron Bombardment Ion Engines, *J. Spacecraft Rockets* **7,** 266 (1970).

R. D. Fowler and G. E. Gibson, The Production of Intense Beams of Positive Ions, *Phys. Rev.* **46,** 1075 (1934).

M. Fumelli and R. Becherer, *Periplasmatron Ion Sources*, Report EUR-CEA-FC-901, Association Euraton-CEA, 92260 Fontenay-aux-Roses, France, June 1977.

M. Fumelli and F. P. G. Valckx, *Nucl. Instrum. Methods* **135,** 203 (1976).

J. P. Gauyacq and J. J. C. Geerlings, H⁻ Formation in H⁺ Surface Collisions, Palaiseau 1986.*

J. J. C. Geerlings, R. Rodink, P. W. van Amersfoort, J. Los, and H. J. Hopman, Formation of Light Negative Ions in Low Work Function Surfaces, Palaiseau 1986.

R. Geller, B. Jacquot, C. Jacquot, and P. Sermot, Project of a New Type of Neutral Injector Based on Negative Deuterons, *Nucl. Instrum. Methods* **175,** 261 (1980).

A. A. Glazov, M. Kuzmyak, D. L. Novikov, and L. M. Onishchenko, Ion Source for a 1-MeV Proton Accelerator, *Prib. Tekh. Eksp.,* No. 1, pp. 34–37, Jan.–Feb. 1964. In English: *Instrum. Exp. Tech.,* Pt. 1, 29 (1964).

D. M. Goebel, Ion Source Discharge Performance and Stability, *Phys. Fluids,* **25,** 1093 (1982).

D. M. Goebel and A. T. Forrester, Rectangular Area Hollow Cathode Ion Sources, 3rd Neutral Beam Heating Workshop, Gatlinburg, Tenn., Oct. 19–23, 1981, paper no. 42.

D. M. Goebel and A. T. Forrester, Plasma Studies on a Hollow Cathode, Magnetic Multipole Ion Source for Neutral Beam Injection, *Rev. Sci. Instrum.* **53,** 810 (1982).

D. M. Goebel, J. T. Crow, and A. T. Forrester, Lanthanum Hexaboride Hollow Cathode for Dense Plasma Production, *Rev. Sci. Instrum.* **49,** 469 (1978).

D. M. Goebel, A. T. Forrester, and S. Johnston, La-Mo Emitters in Hollow Cathodes, *Rev. Sci. Instrum.* **51,** 1468 (1980).

D. M. Goebel, J. T. Crow, and A. T. Forrester, High Current Density Structure, United States Patent No. 4,297,615, Oct. 27, 1981.

D. M. Goebel, Y. Hirooka, and T. A. Sketchley, Large-Area Lanthanum Hexaboride Electron Emitter, *Rev. Sci. Instrum.* **56,** 1717 (1985).

A. P. H. Goede and T. S. Green, Operation Limits of Multipole Ion Sources, *Phys. Fluids* **25,** 1797 (1982).

W. G. Graham, Experimental Measurements of Negative Hydrogen Ion Production from Surfaces, Brookhaven 1977, p. 53.

W. G. Graham, Properties of Alkali Metals Adsorbed into Metal Surfaces, Brookhaven 1980, p. 126.

E. H. A. Granneman, J. J. C. Geerlings, J. N. M. van Wunnik, P. J. van Bommel, H. J. Hopman, and J. Los, H⁻ and Li⁻ Formation by Scattering H₂⁺, H₂⁺, and Li⁺ from Cesiated Tungsten Surfaces, Brookhaven 1983.

J. A. Greer and M. Seidl, Sputtering Yields of Negative Hydrogen Ions, Brookhaven 1983, p. 220.

A. Guthrie and R. K. Wakerling, eds. *The Characteristics of Electrical Discharges in Magnetic Fields,* McGraw-Hill, New York (1949).

---

*Palaiseau 1986 refers to the Proceedings of the Second European Workshop, Palaiseau, France, Mar. 5–7, 1986.

G. W. Hamilton, M. Bacal, A. M. Bruneteau, H. J. Doucet, and M. Nachman, Measurement of H⁻ and D⁻ Density in Plasma by Photodetachment, Brookhaven (1980).

J. M. E. Harper, J. J. Cuomo, P. A. Leary, G. M. Summa, H. R. Kaufman, and F. J. Bresnock, Low Energy Ion Beam Etching, *Proc. 9th Int. Conf. Electron and Ion Beam Science and Technology*, vol. 80-6, 1980, p. 518.

E. R. Harrison and W. B. Thompson, The Low Pressure Plane Symmetric Discharge, *Proc. Phys. Soc. (London)* **74**, 145 (1959).

S. C. Haydon, *An Introduction to Discharge and Plasma Physics*, University of New England Press, Armidale, Australia (1964).

E. B. Henschke and S. E. Derby, Full-Plane Threshold Energies for Cathode Sputtering of Metals with Ar^+ Ions, *J. Appl. Phys.* **34**, 2458 (1963).

A. Hershcovitch and K. Prelec, Mark V Magnetron with H. C. D. Plasma Injection, Brookhaven (1980).

A. Hershcovitch and K. Prelec, Hollow Cathode Discharge as a Plasma Source for H⁻, *Rev. Sci. Instrum.* **52**, 1459 (1981).

A. I. Hershcovitch, V. J. Kovarik, and K. Prelec, High-Intensity H⁻ Ion Source with Steady-State Plasma Injection, *Rev. Sci. Instrum.* **57**, 827 (1986).

N. Hershkowitz, K. N. Leung, and T. Romesser, Plasma Leakage Through a Low-β Line Cusp, *Phys. Rev. Lett.* **35**, 277 (1975).

M. Hirsh and H. Oskam, *Gaseous Electronics*, Academic Press, New York (1978).

J. R. Hiskes, Cross Sections for the Vibrational Excitation of the H_2 ($X^1 \Sigma^+$) State via Electron Collisional Excitation of the Higher Singlet States, *J. Appl. Phys.* **51**, 4592 (1980).

J. R. Hiskes and A. Karo, Formation Processes and Secondary Emission Coefficients for H⁻ Production on Alkali-Coated Surfaces, Brookhaven 1977, p. 42.

J. R. Hiskes, A. Karo, and M. Gardner, Mechanism for Negative-Ion Production in the Surface-Plasma Negative-Hydrogen-Ion Source, *J. Appl. Phys.* **47**, 3889 (1976).

J. C. Holloday, *Picket Fence*, USAEC Report WASH-184 (1954), p. 87.

A. T. J. Holmes, Role of the Anode Area in the Behavior of Magnetic Multipole Discharges, *Rev. Sci. Instrum.* **52**, 1814 (1981).

A. J. T. Holmes and T. S. Green, Extraction and Acceleration of H⁻ Ions from a Magnetic Multipole Source, Brookhaven 1983, p. 429.

E. B. Hooper, Jr., and P. A. Willman, Angular Scattering in Charge-Exchanging Beams, *J. Appl. Phys.* **48**, 1041 (1977).

E. B. Hooper, Jr., P. Poulsen, and P. A. Pincosy, High-Current D⁻ Production by Charge Exchange in Sodium, *J. Appl. Phys.* **52**, 7027 (1981).

F. N. Huffman and P. E. Oettinger, Low Work Function Surface for Improving the Yield of Negative Hydrogen Ions, Brookhaven 1980, p. 119.

A. W. Hull, The Dispenser Cathode. A New Type of Thermionic Cathode for Gaseous Discharge Tubes, *Phys. Rev.* **56**, 86 (1939).

J. Hyman, Jr., W. O. Eckhardt, R. C. Knechtli, and C. R. Buckey, Formation of Ion Beams from Plasma Sources: Part I, AIAA J. **2**, 1739 (1964).

D. L. Jacobson and E. K. Storms, Work Function Measurements of Lanthanum-Boron Compounds, *IEEE Trans. Plasma Sci.* **PS-6**, 191 (1978).

R. G. Jahn, *Physics of Electric Propulsion*, McGraw-Hill, New York (1968).

E. James, R. Worlock, T. Dillon, G. Gant, L. Jan, and G. Trump, A One Millipound Cesium Ion Thruster System, AIAA Electric Propulsion Conference, Stanford, Calif., Aug. 31–Sept. 2, 1970, paper no. 70-1149.

R. O. Jenkins and W. G. Trodden, *Electron and Ion Emission from Solids,* Dover, New York (1965).

K. Jimbo, K. W. Ehlers, K. N. Leung, and R. V. Pyle, Volume Production of Negative Hydrogen and Deuterium Ions in a Reflex-Type Ion Source, *Nucl. Instrum Methods* **A248,** 282 (1986).

H. R. Jory and A. W. Trivelpiece, Exact Relativistic Solution for the One-Dimensional Diode, *J. Appl. Phys.* **40,** 3924 (1969).

H. R. Kaufman, Technology of Electron-Bombardment Ion Thrusters, *Adv. Electron. Electron Phys.* **36,** 265 (1974).

H. R. Kaufman and P. D. Reader, Experimental Performance of Ion Rockets Employing Electron-Bombardment Ion Sources, ARS Electrostatic Conference, Monterey, Calif., Nov. 3–4, 1960. Reproduced in Langmuir et al. (1961) and Forrester and Kuskevics (1968).

G. G. Kelley, N. H. Lazar, and O. B. Morgan, A Source for the Production of Large DC Ion Currents, *Nucl. Instrum. Methods* **10,** 263 (1961).

L. J. Kieffer, Low-Energy Electron-Collision Cross-Section Data, Part I: Ionization, Dissociation, Vibrational Excitation, *Atomic Data* **1,** 19 (1969).

J. Kim, J. H. Whealton, and G. Schilling, A Study of Two-Stage Ion Beam Optics, *J. Appl. Phys.* **49,** 517 (1978).

H. J. King, W. O. Eckhardt, J. W. Ward, and R. C. Knechtli, Electron-Bombardment Thrusters Using Liquid-Mercury Cathodes, *J. Spacecraft Rockets* **4,** 599 (1967). Reproduced in Forrester and Kuskevics (1968).

P. T. Kirstein, G. S. Kino, and W. E. Waters, *Space-Charge Flow*, McGraw-Hill, New York (1967).

J. Kistemaker, Potential Distribution in Magnetic Arc Discharges, *Appl. Sci. Res. B* **5,** 313 (1955).

J. Kistemaker, P. K. Rol, J. Schutten, and C. de Vries, The Ion Source of the Amsterdam Isotope Separator, *Z. Naturforsch.,* **10a,** 850 (1955).

C. Kittel, *Introduction to Solid State Physics,* 3rd ed., Wiley, New York (1966).

W. Knauer, R. L. Poeschel, and J. W. Ward, The Radial Field Kaufman Thruster, AIAA 7th Electric Propulsion Conference, Williamsburg, Va., March 3–5, 1969, paper no. 69-259.

M. Kobayashi, K. Prelec, and Th. Sluyters, Studies of the Hollow Cathode Duoplasmatron as a Source of H⁻ Ions, *Rev. Sci. Instrum.* **47,** 1425 (1976).

W. H. Kohl, *Handbook of Materials and Techniques for Vacuum Devices,* Reinhold, New York (1967).

V. J. Kovarik, A. I. Hershcovitch, and K. Prelec, Initiation of Hot Cathode Arc Discharges by Electron Confinement in Penning and Magnetron Configurations, *Rev. Sci. Instrum.* **53,** 819 (1982).

G. Labaune, *Supersonic Thin Vapor Target for Production of D⁻ Ion Beams,* Doctoral dissertation, University of Paris South at Orsay, June 1978. (in French)

J. M. Lafferty, Boride Cathodes, *J. Appl. Phys.* **22,** 299 (1951).

N. D. Lang, Ionization Probability of Sputtered Atoms, *Phys. Rev. B* **27,** 2019 (1983).

D. B. Langmuir, Theoretical Limitations of Cathode Ray Tubes, *Proc. IRE* **25,** 977 (1937).

D. B. Langmuir, E. Stuhlinger, and J. M. Sellen, Jr., eds., *Electrostatic Propulsion,* Academic Press, New York (1961).

D. B. Langmuir, E. Stuhlinger, and J. M. Sellen, Jr., eds., *Progress in Astronautics and Rocketry: Electrostatic Propulsion,* Vol. 5, Academic Press, New York (1961).

I. Langmuir, The Effect of Space Charge and Residual Gases on Thermionic Currents in High Vacuum, *Phys. Rev.* **2,** 450 (1913). Reprinted in Suits (1961), Vol. 3, p. 3.

I. Langmuir, The Effect of Space Charge and Initial Velocities on the Potential Distribution and Thermionic Current between Parallel Plane Electrodes, *Phys. Rev.* **21,** 419 (1923). Reprinted in Suits (1961), Vol. 3, p. 95.

I. Langmuir, Scattering of Electrons in Ionized Gases, *Phys. Rev.* **26,** 585 (1925).

I. Langmuir, The Interaction of Electron and Positive Ion Space Charges in Cathode Sheaths, *Phys. Rev.* **33,** 954 (1929). Reprinted in Suits (1961), Vol. 5, p. 140.

I. Langmuir, Electric Discharges in Gases at Low Pressures, *J. Franklin Institute* **214**(3) (1932). Reprinted in Suits (1961), Vol. 4, p. 163.

I. Langmuir, *Phenomena, Atoms and Molecules*, Philosophical Library, New York (1950).

I. Langmuir and K. B. Blodgett, Currents Limited by Space Charge between Coaxial Cylinders, *Phys. Rev.* **22,** 347 (1923). Reprinted in Suits (1961), Vol. 3, p. 115.

I. Langmuir and K. B. Blodgett, Currents Limited by Space Charge between Concentric Spheres, *Phys. Rev.* **23,** 49 (1924). Reprinted in Suits (1961), Vol. 3, p. 125.

I. Langmuir and K. H. Kingdon, Thermionic Effects Caused by Vapours of Alkali Metals, *Proc. Roy Soc. A.* **107,** 61 (1925).

I. Langmuir and J. B. Taylor, The Mobility of Caesium Atoms Adsorbed on Tungsten, *Phys. Rev.* **40,** 463 (1932). Reprinted in Suits (1961), Vol. 3, p. 330, and in Langmuir (1950), Chap. 17.

G. P. Lawrence, R. K. Beauchamp, and J. L. McKibben, Direct Extraction of Negative Ion Beams of Good Intensity from a Duoplasmatron, *Nucl. Instrum. Methods* **32,** 356 (1965).

J. D. Lawson, *The Physics of Charged Particle Beams,* Oxford University Press, Oxford, England (1977).

C. Lejeune and J. Aubert, Emittance and Brightness: Definitions and Measurements, *Adv. Electron. Electron Phys.,* Supplement 13A, 159 (1980).

K. N. Leung, UCLA Seminar, Feb. 5, 1987. To be published.

K. N. Leung and K. W. Ehlers, Self-Extraction Negative Ion Source, *Rev. Sci. Instrum.* **53,** 803 (1982).

K. N. Leung and K. W. Ehlers, H^- Production from Different Metallic Converter Surfaces, Brookhaven 1983, p. 265.

K. N. Leung and K. W. Ehlers, Volume H^- Ion Production Experiments at LBL, Brookhaven 1983, p. 67.

K. N. Leung, K. W. Ehlers, and M. Bacal, Extraction of Volume-Produced H^- Ions from a Multicusp Source, *Rev. Sci. Instrum.* **54,** 56 (1983).

K. N. Leung, K. W. Ehlers, and R. V. Pyle, Effect of Wall Material on H^- Production in a Multicusp Source, *Appl. Phys. Lett.* **47,** 227 (1985).

K. N. Leung, C. J. DeVries, K. W. Ehlers, L. T. Jackson, J. W. Stearns, M. D. Williams, M. G. McHarg, D. P. Ball, W. T. Lewis, and P. W. Allison, Operation of a Dudnikov Type Penning Source with LaB_6 Cathodes, Brookhaven, 1986.

K. N. Leung, D. Moussa, and S. B. Wilde, Directly Heated Lanthanum Hexaboride Cathode, *Rev. Sci. Instrum.* **57,** 1274 (1986).

R. Levi, New Dispenser Type Cathode, *J. Appl. Phys.* **24,** 233 (1953). and R. Levi, Improved Impregnated Cathode, *J. Appl. Phys.* **26,** 639 (1955).

H. Levine, Department of Mathematics, Stanford University. Private communication, May 30, 1984.

L. M. Lidsky, S. D. Rothleder, D. J. Rose, S. Yoshikawa, C. Michelson, and R. J. Mackin, Jr., Highly Ionized Hollow Cathode Discharge, *J. Appl. Phys.* **33,** 2490 (1962).

R. Limpaecher and K. R. MacKenzie, Magnetic Multipole Containment of Large Uniform Collisionless Quiescent Plasmas, *Rev. Sci. Instrum.* **44,** 726 (1973).

J. F. Mahoney, J. Perel, and A. T. Forrester, Capillaritron: A New, Versatile Ion Source, *Appl. Phys. Lett.* **38,** 320 (1981).

R. A. Mapleton, *Theory of Charge Exchange,* Wiley-Interscience, New York (1972).

F. E. Marble and J. Surugue, eds., Physics and Technology of Ion Motors, *Proceedings of a Technical Meeting of the Combustion and Propulsion Panel at the 13th AGARD General Assembly, Athens Greece, July 15–17, 1963,* Gordon and Breach, New York (1966).

B. Marchant, G. Kuskevics, and A. T. Forrester, *Surface Ionization Microscope,* AIAA Electric Propulsion Conference, Colorado Springs, Colo. Mar. 11–13, 1963, paper no. 63-018.

P. M. Margosian, Preliminary Tests of Insulated Accelerator Grid for Electron-Bombardment Thrustor, NASA Report No. TM-X-1342 National Aeronautics and Space Administration, Lewis Research Center, Cleveland, Ohio (1967).

L. L. Marino, A. C. Smith, and E. Coplinger, Charge Transfer Between Positive Cesium Ions and Cesium Atoms, *Phys. Rev.* **128,** 2243 (1962).

T. D. Masek, Plasma Properties and Performance of Mercury Ion Thrusters, AIAA 7th Electric Propulsion Conference, Williamsburg, Va., March 3–5, 1969. AIA Paper No. 69-256.

T. D. Masek, Plasma Properties and Performance of Mercury Ion Thrusters, *AIAA J.* **9,** 205 (1971).

H. S. W. Massey, *Negative Ions,* 3rd ed., Cambridge University Press, Cambridge (1976).

R. McAdams, A. J. T. Holmes, M. P. S. Nightingale, L. M. Lea, M. D. Hinton, A. F. Newman, and T. S. Green, Production and Formation of Intense H$^-$ Beams, Brookhaven 1986.

R. H. McFarland, A. S. Schlachter, J. W. Stearns, B. Liu, and R. E. Olson, D$^-$ Production by Charge Transfer of 0.3–3 keV D$^+$ in Thick Alkaline-Earth Vapor Targets: Interaction Energies for CaH$^+$, CaH, and CaH$^-$, *Phys. Rev. A* **26,** 775 (1982).

M. M. Menon, C. C. Tsai, J. H. Whealton, D. E. Schechter, G. C. Barber, S. K. Combs, W. K. Dagenhart, W. L. Gardner, H. H. Haselton, N. S. Ponte, P. M. Ryan, W. L. Stirling, and R. E. Wright, Quasi-Steady-State Multimegawatt Ion Source for Neutral Beam Injection, *Rev. Sci. Instrum.* **56,** 242 (1985).

R. X. Meyer, A Space-Charge-Sheath Electric Thruster, *AIAA J.* **5,** 2057 (1967).

R. X. Meyer, Laboratory Testing of the Space-Charge-Sheath Electric Thruster Concept, *J. Spacecraft Rockets* **7,** 251 (1970).

R. B. Miller, *An Introduction to the Physics of Intense Charged Particle Beams,* Plenum Press, New York (1982).

R. D. Moore, *Magneto-Electrostatically Contained Plasma Ion Thruster,* AIAA 7th Electric Propulsion Conference, Williamsburg, Va., March 3–5, 1969, paper no. 69-260.

S. Nakanishi, E. A. Richley, and B. A. Banks, High Perveance Accelerator Grids for Low-Voltage Kaufman Thrusters, *J. Spacecraft Rockets* **5,** 356 (1968).

G. M. Nazarian and H. Shelton, Theory of Ion Emission from Porous Media, American Rocket Society Electrostatic Propulsion Conference, Monterey, Calif., Nov. 3–4, 1960. Reprinted in Langmuir, Stuhlinger, and Sellen (1961), p. 91.

W. B. Nottingham, Ionization and Excitation in Mercury Vapor Produced by Electron Bombardment, *Phys. Rev.* **55,** 203 (1939).

W. B. Nottingham, Thermionic Emission, *Handb. Phys.* **21** (1956); also published as MIT Research Laboratory of Electronics Technical Report 321, Dec. 10, 1956.

K. J. Nygaard, Electron Impact Ionization Cross Section in Cesium, *J. Chem. Phys.* **49,** 1995 (1968). See also Barnett et al. (1977), C.4.30 and C.4.31.

Y. Oka and T. Kuroda, Effects of a Multipole Magnetic Field on Characteristics of a Multifilament Plasma Source, *Appl. Phys. Lett.* **34,** 134 (1979).

Y. Okumura, H. Horiike, T. Inoue, T. Kurashima, S. Matsuda, Y. Ohara, and S. Tanaka, A High Current Volume H⁻ Ion Source with Multi-Aperture Extraction, Brookhaven 1986.

M. L. E. Oliphant and E. Rutherford, Experiments on the Transmutation of Elements by Protons, *Proc. Roy. Soc. A* **141,** 259 (1933).

R. E. Olson, H⁻ Formation Processes of Interest to Ion Source Development, *IEEE Trans. Nucl. Sci.* **NS-23,** 971 (1976).

J. E. Osher, F. J. Gordon, and G. W. Hamilton, Production of Intense Negative Ion Beams, Proc. 2nd Int. Ion Source Conference, Vienna, Austria, Sept. 11–16, 1972.

R. L. Palmer, Observance of H⁻ by Surface Chemi-Ionization on W(110), Brookhaven 1983, p. 281.

C. A. Papageorgopoulos and J. M. Chen, LEED and Work Function Study of Cs and O Adsorption on W(112), *J. Vac. Sci. Techn.* **9,** 570 (1972).

C. A. Papageorgopoulos and J. M. Chen, Coadsorption of Electropositive and Electronegative Elements I, Cs and H_2 on W(100), *Surf. Sci.* **39,** 283 (1973).

J. Pelletier and C. Pomot, Work Function of Sintered Lanthanum Hexaboride, *Appl. Phys. Lett.* **34,** 249 (1979).

F. M. Penning, A New Manometer for Pressures between 10^{-3} and 10^{-5} Torr, *Physica* **4,** 71 (1937).

F. M. Penning and J. H. A. Moubis, A Neutron Source Without Extra Pumping, *Physica* **4,** 1190 (1937).

J. Perel, Alkali Metal Ion Sources. *J. Electrochem. Soc.* **115,** 343C (1968).

J. Perel, R. H. Vernon, and H. L. Daley, Measurement of Cesium and Rubidium Change-Transfer Cross Sections, *Phys. Rev.* **138A,** 937 (1965).

J. Perel, M. F. Mahoney, B. E. Kalensher, and A. T. Forrester, Investigation of the Capillaritron Ion Source for Electric Propulsion, AIAA 15th Int. Electric Propulsion Conference, April 21–23, 1981, Las Vegas, Nev., paper no. 81-0747.

J. R. Pierce, Rectilinear Electron Flow in Beams, *J. Appl. Phys.* **11,** 548 (1940).

J. R. Pierce, *Theory and Design of Electron Beams,* Van Nostrand, New York (1954).

P. A. Pincosy and K. N. Leung, Shaped LaB_6 Filaments for Ion Source Operations, *Bull. Am. Phys. Soc.* **29,** 1312 (1984).

K. Prelec, Progress in the Development of High Current, Steady State H⁻/D⁻ Sources at BNL, Brookhaven 1980, p. 145.

K. Prelec and Th. Sluyters, Formation of Negative Hydrogen Ions in Direct Extraction Sources, *Rev. Sci. Instrum.* **44,** 1451 (1973).

D. C. Prince, Mercury Arc Rectifier Phenomena, *AIEE Journal,* **46,** 667 (1927).

W. Ramsey, 12 Centimeter Magneto-electrostatic Containment Mercury Ion Thruster Development, AIAA/SAE Propulsion Conference, Salt Lake City, Utah, June, 14–18, 1971, paper no. 71-692.

V. K. Rawlin and E. V. Pawlik, A Mercury Plasma-Bridge Neutralizer, *J. Spacecraft Rockets* **5,** 814 (1968).

J. Sakuraba, M. Akiba, Y. Arakawa, H. Horiike, M. Kawai, Y. Ohara, and S. Tanaka, Preliminary Experiments on the Lambatron Ion Source for the JT-60 Neutral Beam Injector, *Rev. Sci. Instrum.* **52,** 689 (1981).

B. Salzberg and A. V. Haeff, Effects of Space Charge in the Grid-Anode Region of Vacuum Tubes, *RCA Rev.* **2,** 336 (1938).

D. E. Schechter and C. C. Tsai, Indirectly Heated Cathodes and Duoplasmation-Type Electron Feeds for Positive Ion Sources, Proc. 3rd Neutral Beam Heating Workshop, Gatlinburg, Tenn. Oct. 19–23, 1981, paper A1.

A. S. Schlachter and T. J. Morgan, Formation of H⁻ by Charge Transfer in Alkaline Earth Vapors, Brookhaven 1983, p. 149.

A. S. Schlachter, K. R. Stalder, and J. W. Stearns, D⁻ Production by Charge Transfer of 0.3–10 keV D⁺, D° and D⁻ in Cesium, Rubidium and Sodium Vapor Targets, *Phys. Rev. A* **22,** 2494 (1980).

P. J. Schneider, K. H. Berkner, W. A. Graham, R. V. Pyle, and J. W. Stearns, D⁻ Production by Backscattering from Clean Alkali-Metal Surfaces, Brookhaven 1977, p. 63.

W. Schottky, Cold and Hot Electron Discharges (in German), *Z. Physik.,* **14,** 63 (1923).

B. W. Schumacher and J. H. Fink, Beam Terminology, *IEEE Trans. Nucl. Sci.* **NS-3,** 403 (1973).

F. Seitz, *The Modern Theory of Solids,* McGraw-Hill, New York (1940).

W. S. Seitz and S. L. Eilenberg, Numerical Self-Consistent Field Approximation to the Interaction of an Ion Beam with a Plasma Boundary, *J. Appl. Phys.* **38,** 276 (1967).

S. A. Self, Exact Solution of the Collisionless Plasma-Sheath Equation, *Phys. Fluids* **6,** 1762 (1963).

N. N. Semashko, V. V. Kusnetov, and A. I. Krylov, Production of Negative Hydrogen Ion Beams by Double Charge Exchange, Brookhaven 1977, p. 170.

N. N. Semashko, V. V. Kuznetsov, A. I. Krylov, and P. S. Firsov, Negative Ion Production by Double Charge Exchange Technique and Their Additional Acceleration, Brookhaven 1986.

H. Shelton, R. F. Wuerker, and J. M. Sellen, American Rocket Society meeting, San Diego, Calif., June 8–11, 1959, paper no. 882-59.

Th. Sluyters, Negative Hydrogen Sources for Beam Currents between One Milliampere and One Ampere, Proc. 2nd Symp. Ion Sources and Formation of Ion Beams, Berkeley, Calif., Oct. 22–25, 1974, paper VIII-2.

Th. Sluyters and K. Prelec, A Hollow Discharge Duoplasmatron as a Negative Hydrogen Ion Source, *Nucl. Instrum. Methods* **113,** 299 (1973).

B. M. Smirnov and M. I. Chibisov, Resonance Charge Transfer in Inert Gases, *Sov. Phys. Tech. Phys.* **10,** 88 (1965).

F. J. Smith, Oscillations in the Caesium and Rubidium Resonant Charge Transfer Cross Sections, *Phys. Lett.* **20,** 271 (1966).

H. V. Smith, Jr., P. Allison, and J. D. Sherman, A Rotating Penning Surface-Plasma Source for DC H⁻ Beams, Brookhaven 1980, p. 178.

L. P. Smith and P. L. Hartman, The Formation and Maintenance of Electron and Ion Beams, *J. Appl. Phys.,* **11,** 220 (1940).

G. Sohl and V. V. Fosnight, Thrust Vectoring of Ion Engines, *J. Spacecraft Rockets* **6,** 143 (1969).

G. Sohl, A. C. Reid, and R. C. Speiser, Cesium Electron Bombardment Ion Engines, *J. Spacecraft Rockets* **3,** 1093 (1966). Reprinted in Forrester and Kuskevics (1968).

J. S. Sovey, Performance of a Magnetic Multipole Line-Cusp Argon Ion Thruster, AIAA Electric Propulsion Conference, Las Vegas, Nevada, April 21–23, 1981, paper no. 81-0745.

R. C. Speiser, Technology and Development of Bombardment Ion Engines, in Marble and Surugue (1966), p. 255.

L. Spitzer, Jr., *Physics of Fully Ionized Gases*, 2nd ed. rev., Interscience, New York (1962).

W. L. Stirling, C. C. Tsai, and P. M. Ryan, 15 cm DuoPIGatron Ion Source, *Rev. Sci. Instrum.* **48**, 533 (1977).

E. Stuhlinger, *Ion Propulsion for Space Flight,* McGraw-Hill, New York (1964).

R. N. Sudan and R. V. Lovelace, Generation of Intense Ion Beams in Pulsed Diodes, *Phys. Rev. Lett.* **31**, 1174 (1973).

C. Guy Suits, ed., *The Collected Works of I. Langmuir,* Vols. 3, 4, 5. Pergamon Press, New York (1961).

J. B. Taylor and I. Langmuir, The Evaporation of Atoms, Ions and Electrons from Caesium Films on Tungsten, *Phys. Rev.* **44**, 432 (1933). Reprinted in Suits (1961), Vol. 3, p. 376, and I. Langmuir (1950), Chap. 16.

J. J. Thomson, Rays of Positive Electricity, *Phil. Mag.* **20**, 752 (1910).

J. J. Thomson, Rays of Positive Electricity, *Phil. Mag.* **21**, 225 (1911).

L. Tonks and I. Langmuir, General Theory of the Plasma of an Arc, *Phys. Rev.* **34**, 876 (1929). Reprinted in Suits (1961), Vol. 5, p. 176.

J. L. Tuck, *Picket Fence,* USAEC Report WASH-184 (1954), p. 77.

M. A. Tuve, O. Dahl, and L. R. Hafstad, The Production and Focusing of Intense Positive Ion Beams, *Phys. Rev.* **48**, 241 (1935).

L. Valyi, *Atom and Ion Sources,* Wiley, New York (1977).

J. P. VanDevender and D. L. Cook, Inertial Confinement Fusion with Light Ion Beams, *Science* **232**, 831 (1986).

R. K. Wakerling and A. Guthrie, eds., *Electromagnetic Separation of Isotopes in Commercial Quantities,* U.C. Radiation Laboratory, TID-5217, June 1949.

J. H. Whealton, Transverse Emittance Compressor, *Bull. Am. Phys. Soc.* **29**, 1262 (1984).

J. H. Whealton, Review of Computer Modeling of Negative Ion Beam Formation, Brookhaven 1986.

J. H. Whealton, L. R. Grisham, C. C. Tsai, and W. L. Stirling, Effect of Preacceleration Voltage upon Ion Beam Divergence, *J. Appl. Phys.* **49**, 3091 (1978).

J. H. Whealton, C. C. Tsai, W. K. Dagenhart, W. L. Gardner, H. H. Haselton, J. Kim, M. M. Menon, P. M. Ryan, D. E. Schechter, and W. L. Stirling, Effect of Preacceleration on Intense Ion-Beam Transmission Efficiency, *Appl. Phys. Lett.* **33**, 278 (1978).

K. Wiesmann, Some Physical Aspects of H⁻ Sources, Brookhaven 1977, p. 97.

R. G. Wilson, Vacuum Thermionic Work Functions of Polycrystalline Nb, Mo, Ta, W, Re, Os and Ir, *J. Appl. Phys.* **37**, 3170 (1966).

R. G. Wilson, Surface Ionization of Indium and Aluminum on Iridium, *J. Appl. Phys.* **44**, 2130 (1973).

R. G. Wilson and G. R. Brewer, *Ion Beams with Applications to Ion Implantation,* Wiley, New York (1973).

R. L. Witkover, Operational Experience with the BNL Magnetron H⁻ Source, Brookhaven 1983, p. 398.

M. L. Yu, Work Function Dependence and Isotope Effect in the Production of Negative Hydrogen Ions During Sputtering of Adsorbed Hydrogen on Cs Covered Mo (100) Surfaces, Brookhaven 1977, p. 48.

I. P. Zapesochnyi and I. S. Aleksakhin, Soviet Physics, *JETP Lett.* **28**, 41 (1969). See also Barnett et al. (1977), C.4.30 and C.4.31.

W. H. Zinn, Low Voltage Positive Ion Source, *Phys. Rev.* **52**, 665 (1937).

D. Zuccaro, R. C. Speiser, and J. M. Teem, Characteristics of Porous Surface Ionizers, American Rocket Society Electrostatic Propulsion Conference, Monterey, Calif., Nov. 3–4, 1960. Reprinted in Langmuir, Stuhlinger, and Sellen (1961), p. 107.

Appendix

Problem Solutions

A Comment on the Grouping of Constants

The following comment on solving numerical problems may be of some use. The units used in this book, unless otherwise specified, are SI or mks units. (Answers to calculations are often expressed in convenient laboratory-sale mixed units such as milliampere or centimeters.) In applying these units certain combinations occur routinely and one can minimize the number of times it is necessary to consult tables of physical constants by some reorganization of the terms. For example, a frequently encountered combination of constants is kT/M where k is the Boltzmann constant, T is a temperature in kelvins, and M the mass in kilograms. We may write

$$\frac{kT}{M} = \frac{kT}{e} \frac{e}{m} \frac{m}{M_p} \frac{1}{M_A}$$

where e and m are the electron charge and mass, respectively, M_p is the proton mass, and M_A the ion mass in atomic mass units. It appears that we have complicated a simple term but (kT/e) is the temperature in electron volts, the unit usually used in plasma physics. $e/m = 1.76 \times 10^{11}$ C/kg [coulombs/kilogram] is a number so often used that it will move into your memory bank along with $M_p/m = 1836$. M_A is usually known to the nearest integer.

If T is given in kelvins, then it is necessary to recall that 1 eV is equivalent to 11,600 K, leading to

$$\frac{kT}{e} = \frac{T}{11,600}.$$

It also happens that ε_0, the permittivity of free space, usually occurs in combination with the electronic charge and it is helpful to remember that $e/\varepsilon_0 = 1.81 \times 10^{-8}$ volt-meters.

Consider an example. Find the time required for a 100 volt He^+ ion to move a Debye shielding distance in a plasma of density 10^{17} m^{-3} and electron temperature 5 eV.

$$t = \text{distance}/\text{velocity}$$

$$= (\varepsilon_0 \, kT/ne^2)^{1/2}(M/2eV)^{1/2}$$

$$= \left[\frac{1}{n}\frac{kT/e}{e/\varepsilon_0}\frac{M_A M_P/m}{2Ve/m}\right]^{1/2}$$

$$= \left[\frac{(5)(4)(1836)}{10^{17}(1.81 \times 10^{-8})(2)(100)(1.76 \times 10^{11})}\right]$$

$$= 7.59 \times 10^{-10} \text{ second.}$$

I believe that you will find this approach helpful.

Solutions to Problems

2.1 $J = \chi V^{3/2}/a^2$; $\chi = 4.72 \times 10^{-9}$ A$/V^{3/2}$

$= (4.72)(10^{-9}) \, 10^{3/2}/(5 \times 10^{-3})^2 = 5.97$ mA/cm^2.

2.2 $P = JV = \chi V^{5/2}/a^2$; $\chi = 2.33 \times 10^{-6}$ A$/V^{3/2}$

$V^{5/2} = (100)(.1)^2/(2.33)(10^{-6})$; $V = 179$ volts

$J = 100/179 = 0.558$ A/cm^2.

2.3   The Child equation distance $\sqrt{\chi/J} \, V^{3/4}$ corresponding to 3×10^{-3} A/cm^2 of 100 eV A^+ ions is 0.054 cm, so small compared to the 10 cm spacing that virtually all the current returns to the injection plane. For the current $2J_0$ the distance is $0.054/\sqrt{2} = 0.038$ cm. At this distance the potential goes to zero and a virtual emitter exists. The current which flows to the collector is given by $J = \chi V^{3/2}/a^2$ where $a = 9.96$ cm. We obtain

$$J = 8.68 \times 10^{-8} \text{ A/cm}^2,$$

so small compared to J_0 that it justifies the use of $2J_0$ for the current in the region near the injection plane.

2.4   In $V'' = -\rho/\varepsilon_0$ let

$$\rho = (\alpha J/v_1) + (1 - \alpha)J/v_2,$$

where

$$v_1 = \sqrt{-2eV/M_1} \quad \text{and} \quad v_2 = \sqrt{-2eV/M_2}$$

($-$ signs because $V \ll 0$ for $+$ ions). We get

$$V'' = (J/\varepsilon_0)\left[\alpha\sqrt{M_1} + (1 - \alpha)\sqrt{M_2}\right]/\sqrt{-2eV}.$$

A comparison with Eq. (2.4) shows that the equivalent mass is given by

$$M_e = \left[\alpha\sqrt{M_1} + (1 - \alpha)\sqrt{M_2}\right]^2.$$

2.5 $i = 2\pi\chi V_0^{3/2}/r_1(-\beta)^2; \quad V_0 = 100, r_1 = 1, (-\beta)^2 = 0.8454$

$i = 2\pi\chi V_0^{3/2}/(0.8454) \text{ A/cm}$

Solve for $V = \left[ir(-\beta)^2/2\pi\chi\right]^{2/3} = V_0\left[(-\beta)^2 r/0.8454\right]^{2/3}.$

r(cm)	$(-\beta)^2$	V	
$2/1.1 = 1.818$	0.0098	7.630	For planar
$2/1.2 = 1.667$	0.03849	17.926	electrodes 1 cm
$2/1.3 = 1.538$	0.08504	28.819	apart
$2/1.4 = 1.429$	0.14856	39.803	
$2/1.5 = 1.333$	0.2282		$V = V_0 x^{4/3}$
$2/1.6 = 1.250$	0.32333		
$2/1.8 = 1.111$	0.5572	50.589	
		61.137	Both are plotted
		81.241	in Fig. A1

2.6 Solve Eq. (2.23) for $(-\alpha)^2$:

$$(-\alpha)^2 = 4\pi\chi V^{3/2}/I = 0.3418.$$

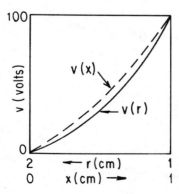

Figure A1. Potential variation in the cylindrical diode of Problem 2.5 compared with $V(x)$ for a planar diode.

Interpolate between $r_0/r = 1.6$ and 1.8 to get $r_0/r = 1.65$, giving

$$r_0 = 1.65 \text{ mm}.$$

2.7 No problem. The electrical connections need exert no forces and power radiated from the system can be radiated uniformly in all directions.

2.8 Rest energy of the electron, $m_0 c^2 = 5 \times 10^5$ eV. Thus at $V = 5 \times 10^6$ volts, $eV/m_0 c^2 = 10$ and we obtain $J/J_0 = 0.61$ from Fig. 2.3.

$$J_0 = (2.33 \times 10^{-6})(5 \times 10^6)^{3/2}/(50)^2 = 10.44 \text{ A/cm}^2$$

so that

$$J = 6.37 \text{ A/cm}^2.$$

2.9 $J = 50 \text{ mA/cm}^2$, H^+ ions $\chi = 5.45 \times 10^{-8} A/V^{3/2}$

$$\eta = V/V_0 = 50/4 = 12.5$$
$$\xi = (\sqrt{\eta} + 2)(\sqrt{\eta} - 1)^{1/2} = 8.814 = a/x_0$$
$$x_0 = \sqrt{\chi/J}\, V_0^{3/4} = 2.95 \times 10^{-3} \text{ cm}$$
$$a = (8.814)(2.95)(10^{-3}) = 2.60 \times 10^{-2} \text{ cm}.$$

From Fig. 2.5, for same energy at collector,

$$a/a_1 = 1.325; \qquad a_1 = 1.965 \times 10^{-2} \text{ cm}$$

and for the same sheath drop,

$$a/a_2 = 1.41; \qquad a_2 = 1.845 \times 10^{-2} \text{ cm}.$$

The two sheath thicknesses a_1 and a_2 can be calculated directly from the Child equation. For 50 volts obtain

$$a_1 = \sqrt{\chi/J}\, V^{3/4} = 1.96 \times 10^{-2} \text{ cm}$$

and for 40 volts

$$a_2 = a_1(46/50)^{3/4} = 1.84 \times 10^{-2} \text{ cm}$$

in perfect agreement.

2.10 (a) $J_0 = \chi V^{3/2}/a^2$; $\chi = 8.61 \times 10^{-9} A/V^{3/2}$
$$= 3.49 \text{ mA/cm}^2$$
(b) $J_1 = 0.75$; $J_0 = 2.62 \text{ mA/cm}^2$

(c) Between the source plasma and grid (Region I)

$$V_0 - V = \left(J_0 x^2 / \chi \right)^{2/3} = 5477 x^{4/3}$$

with x in centimeters and V in volts. $V_0 = 1000$ volts.

Between the grid and the beam plasma (Region II), the situation is identical to that discussed in Sec. 2.5, even though the ions are traveling toward, rather than away from, the $V' = 0$ boundary. We can apply

$$\xi = (\sqrt{\eta} - 1)^{1/2}(\sqrt{\eta} + 2)$$

where $\xi = x_1/x_0$, with x_1 measured from the $V' = 0$ surface back toward the grid and x_0 given by

$$x_0 = \sqrt{\chi/J_1} \, V_0^{3/4} = 0.322 \text{ cm.}$$

The variable $\eta = V/V_0$ varies from 1 to 1.1.

Region I		Region II				
x(cm)	V(volts)	V	η	ξ	x_1	$x = 0.517 - x_1$
0	1000	−10	1.01	0.212	0.068	0.449
0.02	970	−20	1.02	0.300	0.097	0.420
0.05	899	−40	1.04	0.425	0.137	0.380
0.1	746	−60	1.06	0.521	0.168	0.349
0.15	563	−70	1.07	0.563	0.181	0.336
0.2	359	−80	1.08	0.602	0.194	0.323
0.24	137	−90	1.09	0.639	0.206	0.311
0.3	−100	−100	1.1	0.674	0.217	0.3

This tabulation is plotted in Fig. A2.

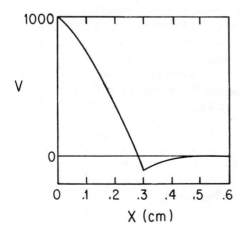

Figure A2. Potential variation between source plasma and beam plasma in Problem 2.10.

2.11 For a current of 9 mA/cm^2 at 1000 volts

$$x_0 = \sqrt{\chi/J}\, V^{3/4} = 0.509 \text{ cm}.$$

Because this is larger than the actual spacing all electrons can get to the zero voltage collector.

For a current $2J$ the Child equation distance becomes

$$x_1 = x_0/\sqrt{2} = 0.36 \text{ cm}.$$

Since this is smaller than the spacing a solution with all the electrons reflected at the distance x_1 is also possible.

2.12 Rising portion of curve: (1) Let α, fraction of current to the collector, be the independent variable. (2) From α get $J = \alpha J_0$. (3) Get $J_1 = (2 - \alpha) J_0$, the space charge equivalent current between insertion grid and virtual cathode. (4) From J_1 and $V_1 = 100$ volts get x_1, distance to the virtual cathode. (5) Get $x_2 = a - x_1$. Finally, from J and x_2 get V_2, the collector voltage.

α	$J(\text{mA/cm}^2)$	$J_1(\text{mA/cm}^2)$	$x_1(\text{mm})$	$x_2(\text{mm})$	V_2
0.05	1.5	58.5	1.997	3.002	14.97
0.1	3	57	2.024	2.976	23.49
0.2	6	54	2.079	2.921	36.37
0.3	9	51	2.139	2.861	46.36
0.4	12	48	2.205	2.795	54.44
0.5	15	45	2.277	2.723	61.01
0.6	18	42	2.357	2.643	66.21
0.7	21	39	2.446	2.554	70.10
0.8	24	36	2.546	2.454	72.65
0.9	27	33	2.659	2.341	73.80
1	30	30	2.789	2.211	73.36

For $V_2 > 73.8$, $V(x)$ will be of the form given by Fig. 2.6c or d. As V_2 is lowered J will remain at 30 mA/cm^2 until $V_0 = V_{0m}$. Approach a solution as follows: (1) Choose a V_2. (2) From V_2/V_1 get V_{0m}/V_1 from Fig. 2.8. (3) From $\eta_1 = V_1/V_{0m}$ and $\eta_2 = V_2/V_{0m}$ get ξ_1 and ξ_2 using Fig. 2.4 or $\xi = (\sqrt{\eta} + 2)(\sqrt{\eta} - 1)^{1/2}$. (4) Compute $x_0 = \sqrt{\chi/J}\, V_0^{3/4}$. (5) Get $x_1 + x_2 = (\xi_1 + \xi_2) x_0$. (6) Repeat until two values of V_2 are found for which $x_1 + x_2$ closely bound the electrode spacing. For example, for $V_2 = 22$ volts, $V_2/V_1 = 0.22$. For $V_2/V_1 = 0.22$ get $V_{0m}/V_1 = 0.1035$, yielding $V_{0m} = 10.35$ volts. Then $\eta_1 = 100/10.35 = 9.662$ and $\eta_2 = 22/10.35 = 2.126$ yielding $\xi_1 = 7.417$ and $\xi_2 = 2.340$. Find $x_0 = 0.05089$ cm leading to $x_1 = 3.775$ cm and $x_2 = 0.1191$ cm leading to $x_1 + x_2 = 0.4996$ cm. For $V_2 = 24$ volts the same procedure leads to $x_1 + x_2 = 0.5072$ cm. Interpolation to a value of V_2 giving $x_1 + x_2 = 0.5$ cm yields $V_2 = 22.6$ volts as the value at which the current must discontinuously drop. Figure A3 is the $J(V)$ curve.

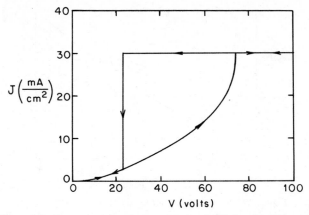

Figure A3. Current–voltage characteristic for the diode of Problem 2.12.

2.13 $J_S = 0.5 \text{ A/cm}^2$, $J = 0.05 \text{ A/cm}^2$, $kT/e = 2320/11{,}600 = 0.2$ volt.
$\eta_c = \ln(0.5/0.05) = 2.3026$
$V_c - V_m = \eta_c kT/e = 0.4605$ volt.
From Fig. 2.13 get

$$\xi_c = -2.15$$

$$\beta = (2J\sqrt{\eta}/9\chi)^{1/2}(kT/e)^{-3/4} = 307.14 \text{ cm}^{-1}$$

$$x_c - x_m = \xi_c/2\beta = -3.5 \times 10^{-3} \text{ cm}$$

$$\xi_a - \xi_c = 2\beta(x_a - x_c) = 0.2\beta = 61.428$$

$$\xi_a = 61.428 - 2.15 = 59.28$$

$$\eta_a = 150$$

$$V_a - V_m = 150 \, kT/e = 30 \text{ volts}$$

$$V_a - V_c = 29.54 \text{ volts}$$

Tabulation of $V(x)$:

$x - x_c$(cm)	$\xi = -2.15 + 2\beta(x - x_c)$	η	$V - V_c = \eta(kT/e) - 0.46$
0.01	3.993	2.5	0.04
0.02	10.136	11.0	1.74
0.03	16.278	23.1	4.16
0.04	22.42	36.9	6.92
0.05	28.56	52.8	10.10
0.06	34.71	72	13.94
0.07	40.85	89	17.34

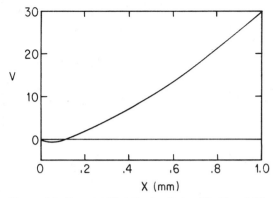

Figure A4. Potential $V(x)$ for the diode of Problem 2.13.

$x - x_c$ (cm)	$\xi = -2.15 + 2\beta(x - x_c)$	η	$V - V_c = \eta(kT/e) - 0.46$
0.08	46.99	110	21.54
0.09	53.14	129	25.34
0.1	59.28	150	29.54

$V - V_c$ as a function of $x - x_c$ is shown in Fig. A4.

2.14 The procedure is indicated by the tabulation following:

J $\left(\dfrac{mA}{cm^2}\right)$	η_c $ln(0.5/J)$	$V_c - V_m$ (volts)	ξ_c (Fig. 2.13)	β (cm^{-1})	ξ_a $\xi_c + 2\beta a$	η_a (Fig. 2.14)	$\dfrac{V_a - V_c}{V_a - V_m}$ $-(V_c - V_m)$
1	6.21	1.24	−2.55	43.4	6.14	5.0	−0.24
5	4.61	0.92	−2.42	97.1	17.01	24.6	4.00
10	3.91	0.78	−2.35	137.4	25.12	43.6	7.94
50	2.30	0.46	−2.10	307	59.3	150	29.54
100	1.61	0.32	−1.85	434	85.0	225	44.68
200	0.92	0.18	−1.60	614	121.3	408	81.42
300	0.51	0.10	−0.60	752	149.9	548	109.5
400	0.22	0.04	−0.30	869	173.4	668	133.6
500	0	0	0	971	194.2	785	157

Let $V_a - V_c = V$, $J(V)$ is plotted in Fig. A5. For the 0.1 cm spacing the Child equation becomes

$$J = 2.334 \times 10^{-4} V^{3/2}$$

where V is in volts and J in amps per square centimeter. This is also plotted in Fig. A5.

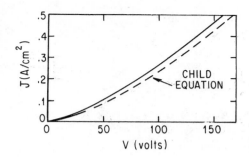

Figure A5. Current–voltage characteristics for the diode of Problem 2.14 compared with the Child equation.

2.15 $V_a - V_c = 10$ volts, with $kT/e = 1000/11{,}600 = 0.0862$ volts yields $\eta_a = \eta_c = 10/0.0862 = 116.0$.

Use this to make a first guess as to the current, as follows. Take $\eta_a = 116$ and use Eq. (2.73) to obtain

$$J_1 = (59.7)(1 + 2.66/\sqrt{116}) = 74.4 \text{ A/m}^2.$$

We can use this to get a value η_c which is fairly accurate, since J_1 is very sensitive to η_c. Obtain

$$\eta_c = \ln \frac{200}{74.4} = 0.99$$

yielding

$$\eta_a = 117$$

Use Fig. 2.13 to obtain $\xi_c = -1.6$ and $\xi_a = 49.5$. Then

$$\xi_a - \xi_c = 2\beta(x_a - x_c)$$

leads to

$$\beta = 51.1/(2)(5 \times 10^{-5}) = 5.11 \times 10^5 \text{ m}^{-1}$$

Solving Eq. (2.42) for J as a function of β yields

$$J = 78.35 \text{ A/m}^2.$$

This is close enough to J_1 to ensure the accuracy of the procedure used. Thus we see the current is 78.35 A/cm^2 rather than 59.7 as calculated in Problem 2.1.

2.16 (a) $J_e/J_i = \sqrt{M/m} = \sqrt{(200)(1936)} = 606$

$J_e = (2 \times 10^{-3})(603) = 1.21 \text{ A/cm}^2$

(b) For a simple Child equation sheath

$$a_0 = (\chi V^{3/2}/J_i)^{1/2}.$$

For $V = 100$ volts, $J_i = 2 \times 10^{-3}$ A/cm^2 and $\chi = 3.85 \times 10^{-9}$ A/V$^{3/2}$

$$a_0 = 0.044 \text{ cm}$$

and

$$a = 1.364 \, a_0 = 0.060 \text{ cm}.$$

2.17 The beam is required to assume the shape shown in Fig. A6. At the beam waist (where the diameter is 0.5 cm) the ions will be traveling parallel to the axis. We may analyze this as for an injected ion beam of diameter 0.5 cm expanding to 2 cm diameter in 20 cm. For an expansion by a factor of 4 we cannot use Fig. 2.16 but must resort to Eq. (2.107). With $r/r_0 = 4$, $\sqrt{\ln{(r/r_0)}} = 1.1774$ my calculator yields

$$\int_0^{1.1774} \exp{(u^2)} \, du = 2.047.$$

With $z/r_0 = 80$, Eq. 2.107 yields

$$\Pi = 18\pi (2.047)^2 / (80)^2 = 0.037$$

leading to a current

$$I = (0.37)(3.85)(10^{-9})(10^6) = 1.425 \times 10^{-4} \text{ A}$$

of 10^4 eV Hg$^+$ ions. To obtain the point z_0 on the axis to which the ions must be aimed note in Fig. A5 that $(r/z_0) = dr/dz$ leading to $z_0 = r \, dz/dr$.

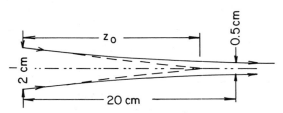

Figure A6. Ion beam configuration for Problem 2.17.

From Eq. (2.107) we obtain

$$\left(\frac{\Pi}{18\pi}\right)^{1/2} \frac{1}{r_0} \frac{dz}{dr} = \frac{d}{dr} \int^{\ln(r/r_0)} \exp{(u^2)} \, du$$

$$= \exp{(\ln r/r_0)} \, d\sqrt{\ln{(r/r_0)}}/dr$$

yielding

$$\frac{dz}{dr} = \frac{\sqrt{18\pi/\Pi}}{2\sqrt{\ln{(r/r_0)}}} = 16.60.$$

Finally we obtain

$$z_0 = r \, dz/dr = 16.60 \text{ cm}$$

since $r = 1$ cm.

3.1 (a) From Eq. (3.9) with $J_0 = \chi V^{3/2}/a^2$ and $\chi = 8.61 \times 10^{-9}$ for A^+ (see Fig. 2.2), we obtain

$$J_1 = (2.865)\,(8.61)\,(10^{-9})\,(5000)^{3/2}/10^4 = 8.72 \times 10^{-7} \text{ A/cm}^2,$$

$$I_1 = 8.72 \ \mu\text{A}.$$

(b) Eq. (3.10) yields

$$\rho = 8.85 \times 10^{-8} \text{ C/m}^3 = 8.85 \times 10^{-14} \text{ C/cm}^3.$$

(c) $S_1 = J_1/ea = (8.72 \times 10^{-7})/(1.6)\,(10^{-19})\,(10^2) = 5.45 \times 10^{10}$ cm^{-3}/s but $S_1 = 0.1 \, n_a \therefore n_a = 5.45 \times 10^{11}$ cm^{-3}.

(d) At 760 torr there are $6.02 \times 10^{23}/2.24 \times 10^4 = 2.69 \times 10^{19}$ atoms/cm^3

$$\therefore p = \frac{5.45 \times 10^{11}}{2.69 \times 10^{19}} (760) = 1.54 \times 10^{-5} \text{ torr.}$$

(e) With a cross section of 3×10^{-16} cm^2 the mean free path would be

$$\lambda = 1/(n\sigma) = 6.12 \times 10^3 \text{ cm} = 6.12 \text{ m.}$$

Yes, it is a very rare electron that would do any ionization.

3.2 Poisson's equation yields

$$\frac{1}{r^2} \frac{d}{dr}\left(r^2 \frac{dV}{dr}\right) = \frac{en_0}{\varepsilon_0}\left(1 - \exp{\frac{eV}{kT}}\right)$$

For $eV/kT \ll 1$ we obtain

$$\frac{1}{r^2} \frac{d}{dr} \left(r^2 \frac{dV}{dr} \right) = \frac{e^2 n_0}{\varepsilon_0 kT} V.$$

Try

$$V = \frac{A}{r} \exp{-\frac{r}{r_0}}$$

and substitute in the differential equation to obtain

$$(A/rr_0^2) \exp{-(r/r_0)} = n_0 e^2 A/\varepsilon_0 kT r \exp{(-r/r_0)}$$

so that the trial function is a solution if

$$r_0 = \lambda_D = (\varepsilon_0 kT/n_0 e^2)^{1/2}.$$

3.3 (a) $\lambda_D = 745 \, [T(eV)/n_0(cm^{-3})]^{1/2} = 745 \, [3000/11,600 \times 10^{18}]^{1/2} = 3.8 \times 10^{-7}$ cm.

(b) $\lambda_D = 745 \, (10^4/10^{15})^{1/2} = 2.36 \times 10^{-3}$ cm.

(c) $\lambda_D = 745 \, (5/10^{12})^{1/2} = 1.67 \times 10^{-3}$ cm.

(d) $\lambda_D = 745 \, (0.5/10)^{1/2} = 167$ cm $= 1.67$ m.

(e) $\lambda_D = 745 \, (1/1) = 7.45$ m.

3.4 From the Bohm criterion

$$v_i = \sqrt{kT_e/M}.$$

From kinetic theory

$$v_{ax} = \sqrt{2kT_a/\pi M}$$

$$v_i/v_{ax} = \sqrt{\pi T_e/2T_a}$$

$$= \sqrt{\pi (5)/(2)(0.05)} = 12.53;$$

$$J_i/J_a = 12.53.$$

3.5 Differentiate all terms with respect to ξ using the identity given in the statement of the problem. Obtain

$$\left(\frac{\lambda_D}{L} \right)^2 \eta' \eta'' + \eta' \epsilon^{-\eta} - \frac{L}{n_0} \sqrt{\frac{2M}{kT}} \int_0^\xi \frac{g(\xi_1) \, \eta'(\xi)}{2\sqrt{\eta - \eta_1}} \, d\xi_1 = 0.$$

Divide through by η' to obtain Eq. (3.35).

3.6 Equation (3.42)

$$\xi = \frac{2}{\pi} \epsilon^{-\eta} \int_0^{\sqrt{\eta}} \exp t^2 \, dt.$$

For small arguments $e^x = 1 + x$ so that

$$\xi = \frac{2}{\pi} (1 - \eta) \int_0^{\sqrt{\eta}} (1 + t^2) \, dt = \frac{2}{\pi} (1 - \eta) \left(\sqrt{\eta} + \frac{1}{3} \eta^{3/2} \right)$$

$$= \frac{2\sqrt{\eta}}{\pi} (1 - \eta) \left(1 + \frac{\eta}{3} \right) = \frac{2\sqrt{\eta}}{\pi} \left(1 - \frac{2\eta}{3} \right).$$

Equation 3.43

$$\xi = \frac{2}{\pi} \left\{ \int_0^{\sqrt{\eta}} (1 + t^2) \, dt - \int_0^{\eta} \left[\int_0^{\sqrt{s}} (1 + t^2) \, dt \right] ds \right\}$$

$$= \frac{2}{\pi} \left[\sqrt{\eta} + \frac{\eta^{3/2}}{3} - \int_0^{\eta} \left(\sqrt{s} + \frac{s^{3/2}}{3} \right) ds \right]$$

$$= \frac{2}{\pi} \left(\sqrt{\eta} + \frac{\eta^{3/2}}{3} - \frac{2}{3} \eta^{3/2} - \frac{2}{15} \eta^{5/2} \right).$$

Neglect the last term to obtain

$$\xi = \frac{2}{\pi} \sqrt{\eta} \left(1 - \frac{\eta}{3} \right).$$

Equation (3.44)

$$\xi = \frac{2}{\pi} \left[\sqrt{2} \int_0^{\sqrt{2\eta}} \exp (t^2) \, dt - \epsilon^{\eta} \int_0^{\sqrt{\eta}} \exp (t^2) \, dt \right]$$

$$= \frac{2}{\pi} \left[\sqrt{2} \int_0^{\sqrt{2\eta}} (1 + t^2) \, dt - (1 + \eta) \int_0^{\sqrt{\eta}} (1 + t^2) \, dt \right]$$

$$= \frac{2}{\pi} \left[2\sqrt{\eta} + \frac{4}{3} \eta^{3/2} - (1 + \eta) \left(\sqrt{\eta} + \frac{\eta^{3/2}}{3} \right) \right]$$

$$= \frac{2}{\pi} \sqrt{\eta}$$

neglecting the term in $\eta^{5/2}$.

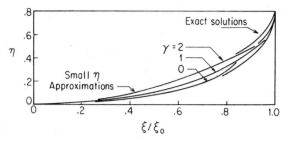

Figure A7. Approximate plasma solutions found in Problem 3.6 compared with the solutions of Harrison and Thompson.

For comparison with Fig. 3.9 these three cases yield

$$\gamma = 0, \; \xi/\xi_0 = 1.848 \sqrt{\eta} \left(1 - \tfrac{2}{3}\eta \right);$$

$$\gamma = 1, \; \xi/\xi_0 = 1.573 \sqrt{\xi} \left(1 - \tfrac{1}{3}\eta \right);$$

$$\gamma = 2, \; \xi/\xi_0 = 1.294 \sqrt{\eta}.$$

These are plotted in Fig. A.7 together with the exact solutions. These approximations are accurate out to values of $\xi/\xi_0 = 0.75$ for $\gamma = 0$, 0.9 for $\gamma = 1$, and 0.8 for $\gamma = 2$.

3.7 $f_\eta(u) = n_0^{-1} \, dn_i/du$, so that $dn_i = n_0 f_\eta(u) \, du$. The contribution of these ions to the current density is given by

$$dJ_i = ve \, dn_i$$

or

$$dJ_i = u \sqrt{2kT/M} \; en_0 \, f_\eta(u) \, du.$$

$$\frac{1}{n_0 e \sqrt{2kT/M}} \frac{dJ_i}{du} = u f_\eta(u)$$

$$= \frac{2u}{\pi} \left[\frac{1}{\sqrt{\eta - u^2}} - 2F(\sqrt{\eta - u^2}) \right]$$

is the desired current distribution function.

3.8 $J_i = \dfrac{(0.344) \, (5 \times 10^{17}) \, (1.6 \times 10^{-19})}{\sqrt{(2)} \, (6) \, (1.76 \times 10^{11})/(1836) \, (131)}.$

3.9 $\lambda_D = 745 \, (5 \times 10^{-10})^{1/2} = 1.666 \times 10^{-2}$ cm. In units of λ_D the distance from the grid to the -30 volts equipotential is $\Delta\zeta = (0.3)/\lambda_D = 18.01$. In Fig. 3.16 start at $\eta = 6$, the value at the desired equipotential, for which the corresponding $\zeta = 36.78$. Subtract $\Delta\zeta$ to get $\zeta_g = 18.77$ for which the corresponding value is $\eta_g = 42.45$. The grid voltage must be

$$V_g = -(42.45)\,(5) = -212.3 \text{ volts.}$$

3.10 From Fig. 3.9 get $n_a = 5.6 \times 10^{16}$ atoms/m^2. With $a = 0.3$ m we obtain $n_a = 1.87 \times 10^{17}$ m^{-3}. For $\eta = 0.9$, $J_i/J_a = 9$ and Eq. (3.73), with $\overline{\sigma v} = 1.52 \times 10^{-14}$ m^3/s (from Fig. 3.8), yields $n_0 = 1.43 \times 10^{17}$ m^{-3}. This leads to an ion current

$$J_i = (3.44)\, n_0 e \sqrt{2kT/M} = 17.2 \text{ A/m}^2 = 1.72 \text{ mA/cm}^2.$$

3.11 From Eq. (3.74) we obtain

$$\frac{J_i}{J_a} = (2)\,(2 \times 10^{18})\,(0.05)\,(0.25)\,(3 \times 10^{-20})$$

$$\cdot \sqrt{\frac{(\pi)\,(1836)\,(40)\,(80)}{0.05}} = 28.8;$$

$$\eta = 28.8/29.8 = 0.966.$$

4.1 $p_0 = (e/\varepsilon_0)/8\pi V = 1.8 \times 10^{-8}/800\pi = 7.16 \times 10^{-12}$ m;

$\lambda_0 = \dfrac{1}{\pi p_0^2 n} = 6.21 \times 10^3$ m;

$v = \sqrt{2eV/m} = 5.93 \times 10^6$ m/s;

$t_0 = \lambda_0/v = 1.05 \times 10^{-3}$ s.

4.2 For collisions with electrons the application of Eqs. (4.34) and (4.40) yields

$$\langle \Delta v_\parallel \rangle_e = -A_D \frac{m}{2kT}\, 2\, \frac{0.5}{\alpha^2} = -A_D \frac{m}{2kT} \frac{2kT}{v^2 m} = -\frac{A_D}{v^2}.$$

For collisions with ions we obtain

$$\langle \Delta v_\parallel \rangle_i = -A_D \frac{M}{2kT_i}\, \frac{0.5}{\alpha^2} = -A_D \frac{M}{4kT_i} \frac{2kT_i}{v^2 M} = -\frac{A_D}{2v^2}.$$

We see that

$$\langle \Delta v_{\parallel} \rangle_e = 2 \langle \Delta v_{\parallel} \rangle_i.$$

4.3 Since $\Phi(\alpha) - G(\alpha) = 1$ for large α, Eq. (4.36) yields

$$\langle (\Delta v_{\perp})^2 \rangle = A_D/v$$

whenever $v \gg \sqrt{2kT/m_f}\ (\alpha \gg 1)$. For large α, Eq. (4.35) becomes

$$\langle (V_{\parallel})^2 \rangle = \frac{0.5}{\alpha^2} \frac{A_D}{v} \ll \langle (\Delta v_{\perp})^2 \rangle.$$

4.4
$$v = \sqrt{2eV/M}; \quad \alpha_i = \sqrt{(2eV/M)(M/2kT)} = \sqrt{eV/kT} \gg 1;$$
$$G(\alpha_i) = kT/2eV = 0.025; \quad \Phi(\alpha_i) - G(\alpha_i) = 1;$$
$$\alpha_e = \sqrt{(eV/kT)(m/M)} \ll 1;$$
$$G(\alpha_e) = (2/3\pi)\,\alpha_e = 6.21 \times 10^{-3}; \quad \Phi - G = 2G.$$

If we take $\ln \Lambda = 10$, we obtain

$$A_D = 2.96 \times 10^{15}\ \mathrm{m}^3/\mathrm{s}^4.$$

From Eq. (4.34) we obtain

$$\langle \Delta v_{\parallel} \rangle_i = -6.18 \times 10^6\ \mathrm{m/s}^2$$

and

$$\langle \Delta v_{\parallel} \rangle_e = -7.67 \times 10^5\ \mathrm{m/s}^2.$$

Slowing of ions by ions is much faster than slowing by electrons. From Eq. (4.35) we obtain

$$\langle (\Delta v_{\parallel})^2 \rangle_i = (3.38)\,(10^9)\ \mathrm{m}^2/\mathrm{s}^3$$

and

$$\langle (\Delta v_{\parallel})^2 \rangle_e = (8.40)\,(10^8)\ \mathrm{m}^2/\mathrm{s}^3.$$

Comparing Eq. (4.36) and (4.35) we see that

$$\langle (\Delta v_{\perp})^2 \rangle = (\Delta v_{\parallel})^2 \frac{\Phi - G}{G},$$

which leads readily to

$$\langle (\Delta v_\perp)^2 \rangle_i = 1.35 \times 10^{11} \text{ m}^2/\text{s}^3$$

and

$$\langle (\Delta v_\perp)^2 \rangle_e = 1.68 \times 10^9 \text{ m}^2/\text{s}^3,$$

showing a deflection rate due to ions which is very much greater.

4.5 We see from Eq. (4.46) that $t_D = t_0/8 \ln \Lambda$. Taking $\ln \Lambda = 10$ we obtain

$$t_D = t_0/80$$

and, using the result of Problem 4.1,

$$t_D = 1.31 \times 10^{-5} \text{ s}.$$

Check: $\ln \Lambda = \ln (\lambda_D/p_0)$. From Eq. (3.20)

$$\lambda_D = 745 \sqrt{2.5 \times 10^{-12}} = 1.18 \times 10^{-3} \text{ cm}$$

and using p_0 from Problem 4.1, we obtain

$$\Lambda = 1.65 \times 10^{-6} \quad \text{and} \quad \ln \Lambda = 14.3.$$

A better value of t_D would be 9.16×10^{-6} s.

5.1 The current density based on the Child equation, $J_C = \chi V^{3/2}/d^2$, where d = electrode separation, yields $I_C = \pi a^2 \chi V^{3/2}/4d^2$, where a = aperture diameter. With $a = d$, $\chi = 1.21 \times 10^{-8}$ A/V$^{3/2}$ (from Fig. 2.2), $V = 10^4$ volts, we obtain

$$I_c = 9.50 \times 10^{-3} \text{ A} = 9.50 \text{ mA}.$$

For the minimum divergence

$$I = 0.4 \, I_c = 3.8 \text{ mA}$$

and

$$\Pi = I/\chi V^{3/2} = 0.31.$$

5.2 The spacing between the plasma electrode and accel electrode is the same as for the Child equation:

$$x_0 = \sqrt{\chi/J} \, V_0^{3/4}.$$

For A^+, Fig. 2.2 gives $\chi = 8.61 \times 10^{-9}$ A/V$^{3/2}$ and with $J = 0.05$ A/cm^2 and $V_0 = 10^4$ we obtain $x_0 = 0.415$ cm. From Fig. 5.5 we see that the divergence angle of an ion at the beam edge is

$$\theta = \frac{a/2}{3x_0/2} = 0.08 \text{ radians} = 4.60°.$$

5.3 $J = \chi V^{3/2}/d^2$ where d = electrode separation yields $I = \pi a^2 \chi V^{3/2}/4d^2$ where a = aperture diameter. With $a = 0.7$ cm, $d = 1.5$ cm, $\chi = 4.72 \times 10^{-9}$ A/V$^{3/2}$ (from Fig. 2.2) we obtain

$$I = 2.85 \times 10^{-4} \text{ A} = 0.285 \text{ mA}$$

and poissance

$$\Pi = I/\chi V^{3/2} = \pi a^2/4d^2 = 0.171.$$

5.4 From the definition of poissance given by Eq. (2.116) we have

$$\Pi = (I/\chi V^{3/2})(a/b).$$

Eq. (2.25) with $r = r_0/4$ and $\beta^2 = (-\beta)^2 = 6$ (from Table 2.1) yields an emitter current density

$$J = 2\chi V^{3/2}/3r_0^2.$$

The cathode area $r_0 \theta b$ yields

$$I = \frac{2\chi V^{3/2}\theta b}{3r_0}.$$

With $a = 2r_0 \sin (\theta/2)$ we obtain

$$\Pi = \frac{4}{3} \theta \sin \frac{\theta}{2}.$$

For $\theta = \pi/6$ radians

$$\Pi = 0.181.$$

5.5 The current density at the emitter is given by Eq. (2.26):

$$J = \chi V^{3/2}/r_0^2 \alpha^2.$$

From Table 2.1 we get $\alpha^2 = (-\alpha)^2 = 5$. The cathode area is $2\pi r_0^2 [1 - \cos(\theta/2)]$. The total emitter current is then

$$I = 0.4\pi\chi V^{3/2}[1 - \cos(\theta/2)]$$

leading to a poissance

$$\Pi = 0.4\pi [1 - \cos(\theta/2)].$$

For $\theta = 30°$ we obtain

$$\Pi = 0.043.$$

5.6   From Eq. (3.56) we obtain the current density

$$J = 0.344\, n_0 e\, \sqrt{2kT/M},$$

leading to

$$J = (0.344)(5 \times 10^{17})(1.6 \times 10^{-19})\sqrt{\frac{(2)(4)(1.76 \times 10^{11})}{(1836)(200)}}$$

$$= 54\ \text{A/m}^2 = 0.0054\ \text{A/cm}^2$$

to the plasma boundary. With a transparency of 0.7 we have an extracted current density of 0.00377 A/cm^2. To obtain 1 A requires an extraction area of 265 cm^2, corresponding to a diameter

$$D = 18.4\ \text{cm}.$$

The current per aperture is limited by perveance considerations to

$$i = 0.25\chi V^{3/2}.$$

From Fig. 2.2 we obtain $\chi = 3.85 \times 10^{-9}\ A/V^{3/2}$ and with $V = 5000$ volts we obtain 3.40×10^{-4} A per aperture. The number of required apertures is therefore

$$N = 2.94 \times 10^3.$$

The aperture area is given by $(3.4 \times 10^{-4}/0.0054)$ cm^2 leading to an aperture diameter

$$d = 0.28\ \text{cm}.$$

5.7 The average current density out of the source of Problem 5.4 is 3.77×10^{-3} A/cm². Under stalling conditions the beam travels out and back so that the effective current, as far as producing space charge, is 7.54×10^{-3} A/cm². Solving the Child equation

$$7.54 \times 10^{-3} = \chi V^{3/2}/a^2$$

for a yields the stalling distance

$$a = 0.42 \text{ cm.}$$

5.8 The best approach here is to use a generalized sheath curve such as Fig. 3.36 and to choose a current such that the difference in distance between $\eta = 60$ and $\eta = 10$ is 0.3 cm. In Fig. 3.36 the corresponding values of ζ are 12.5 and 33.7 giving a $\Delta\zeta = 21.2$, implying that

$$21.2 \, \lambda_D = 0.3 \text{ cm}$$

$$\lambda_D = 0.0142 \text{ cm.}$$

With an electron temperature of 5 eV, Eq. (3.20) yields a plasma density

$$n_0 = 1.38 \times 10^{10} \text{ cm}^{-3}$$

and a current density, from Eq. (3.69), of

$$J = 3.72 \text{ A/m}^2 = 0.372 \text{ mA/cm}^2.$$

5.9 Equation (5.25) yields

$$\theta = \frac{5 \times 10^{-4}}{6} \sqrt{\frac{200}{5.45 \times 10^{-8}}} \, 500^{-3/4}$$

$$= 4.77 \times 10^{-2} \text{ radius} = 2.74°.$$

8.1 Cusp width is given by

$$w = 4\sqrt{r_e r_i} = \frac{4}{eB} \sqrt{mv_e M v_i}$$

$$= \frac{4}{eB} \sqrt{mM} \, (kT_e kT_i / mM)^{1/4}$$

$$= \frac{4}{B\sqrt{e/m}} \left[(M/m)(kT_e e)(kT_i/e) \right]^{1/4}.$$

With $B = 0.32$ tesla, $e/m = 1.76 \times 10^{11}$ C/kg, $M/m = 3672$, $kT_e/e = 8$ volts, and $kT_i/e = 2$ volts we obtain

$$w = 4.64 \times 10^{-4} \text{ m}.$$

Set $12\ wL = \pi r^2$, where $r = 0.075$ m, obtaining $L = 3.17$ m.

8.2 Equation (3.102) gives

$$n_a = \frac{0.34}{\alpha a \sigma_p} \sqrt{\frac{kT}{eV} \frac{m}{M}}.$$

The source half length $a = 1$ m is small enough compared to L in Problem 8.1 that we can ignore the radial ion flow. With $\alpha = 0.05$, $a = 1$ m, $\sigma_p = 9 \times 10^{-21}$ m^2, $kT/e = 8$ volts, $V = 50$ volts and $M/m = 3672$ we obtain

$$n_a = 5.0 \times 10^{18} \text{ m}^{-3}.$$

8.3 $22.4l$ at 760 torr is 1 mole and contains 6.02×10^{23} molecules.

$$1 \text{ torr } l/s = \frac{(6.02)(10^{23})}{(760)(22.4)} = 3.54 \times 10^{19} \text{ molecules/s}$$

$$= 7.07 \times 10^{19} \text{ atoms/s}$$

The equivalent current (assuming 1 electron/atom) is

$$J_{eq} = (7.07)(10^{19})(1.6)(10^{-19}) = 11.3 \text{ A}.$$

The mass utilization efficiency is therefore

$$\eta = 1/11.3 = 0.088, \text{ or } 8.8\%.$$

Index